# Rocks, Gems, and Minerals
of the Southwest

## Help Us Keep This Guide Up to Date

Every effort has been made by the author and editors to make this guide as accurate and useful as possible. However, many things can change after a guide is published.

We would love to hear from you concerning your experiences with this guide and how you feel it could be improved and kept up to date. While we may not be able to respond to all comments and suggestions, we'll take them to heart, and we'll also make certain to share them with the author. Please send your comments and suggestions to the following email address: falconeditorial@rowman.com.

Thanks for your input!

# Rocks, Gems, and Minerals of the Southwest

## SECOND EDITION

Garret Romaine

To my daughter Amy, forever my baby girl.

**FALCONGUIDES®**

An imprint of Globe Pequot, the trade division of
The Rowman & Littlefield Publishing Group, Inc.
4501 Forbes Blvd., Ste. 200
Lanham, MD 20706
www.rowman.com

Falcon and FalconGuides are registered trademarks and
Make Adventure Your Story is a trademark of
The Rowman & Littlefield Publishing Group, Inc.

Distributed by NATIONAL BOOK NETWORK

British Library Cataloguing in Publication Information available

**Library of Congress Cataloging-in-Publication Data**

ISBN 978-1-4930-6441-0 (paper: alk. paper)
ISBN 978-1-4930-6442-7 (electronic)

∞™ The paper used in this publication meets the minimum requirements
of American National Standard for Information Sciences—Permanence of
Paper for Printed Library Materials, ANSI/NISO Z39.48-1992.

# Contents

# Acknowledgments

Special thanks to the staff at the Rice Northwest Museum of Rocks and Minerals in Hillsboro, Oregon. The museum was founded in 1996 and originated with the extensive personal collection of rocks and minerals accumulated by Richard and Helen Rice. Starting with a few beach agates in 1938, the collection is now nationally recognized as one of the finest rock and mineral museums in the United States. The Rice family was especially interested in copper minerals from Arizona and the Southwest, creating a world-class exhibit of azurite, malachite, and more. Also, thanks to the following helpful friends and acquaintances: Rachel Houghton, veteran technical communicator and longtime friend from Portland, Oregon, who helped with photography, touch-up, editing, and encouragement; and Martin Schippers of KleurColor.com in Seattle, Washington, who also helped with photography touch-up.

*Rice Northwest Museum of Rocks and Minerals*

# Introduction

## About Geology

The term "geology" is a combination of two Greek expressions: "Geo" refers to the Earth, and "logos" refers to the logic and language used to explain your observations. So think of geology as a way to organize and explain the Earth processes that we see all around us. There are two key points to consider when trying to understand geology: time and entropy.

- **Time.** The Earth is a very young planet and thus still very active. But it's also pretty old. Even though scientists have measured the Earth at 4.6 billion years old, that's young in the context of a 20-billion-year-old universe. Given enough time, a lot can happen on a young, geologically active planet. We have earthquakes, volcanoes, and moving continents. The forces that boil up from the Earth's magnetic core are a long way from burning out, and they are relentless. Some activities happen quickly, like tsunamis, and they're captured on film. Other forces take millions of years, leaving clues like all the mica flakes lined up in a schist. Good field observers can identify the obvious signs of things that seemed to happen before and apply those signs to the present and future.

- **Entropy.** Things fall apart all the time. Stuff happens. Storms rearrange coastlines and rework river channels. Earthquakes, volcanoes, windstorms, and floods all move mountains and leave scars that "heal." A rock balanced precariously atop another rock will not remain for long; eventually, it will shake loose. The Earth is very efficient at recycling all that surface mayhem, hiding many clues. Mountains rise, then get ground down under glaciers and unrelenting rain. Tight chemical bonds that hold atoms together eventually weaken thanks to water, heat, pressure, and time. Oxygen in the air constantly rusts iron and dissolves minerals. Those forces are always at work and are easy to predict but hard to imagine sometimes. Try to picture the Mississippi River under flooding conditions that happen once every 100,000 years. That's mayhem on a continental scale. Now imagine the resulting gravel bars as the river recedes from flood stage and think about the possibility of being the first rockhound to check those newly stirred gravels.

Given enough time, almost anything can happen, and it usually does. We rarely see these processes at the surface, and we can only imagine what takes place at great depths. That's where the logic comes in. There is a lot of math, chemistry, physics, biology, and just general science involved in sorting out what's going on in the field. But you're mostly interested in what you can see and collect, so read on.

## Think in Series

We don't get a lot of absolutes in nature, so numbers, such as percentages of minerals present, help when thinking about crystal compositions. Just as there are probably no two snowflakes that are exactly alike, most granites differ in some way. Some basalts may have more iron and magnesium present, and some may have more feldspar. You can't exactly tell without expensive equipment. And usually, it doesn't really matter to that many decimals if you have a rough idea. You just want to collect the interesting forms, and you don't need a PhD in structural geology to dig out a seam of agate. You do need a hard hat, however.

Minerals are usually more straightforward than rocks because they tend to be more rigid in their chemistry. Sodium chloride, known as halite or rock salt, has one sodium atom and one chlorine atom. There isn't a lot of wiggle room there. You can get some potassium in the crystal lattice, substituting itself for sodium, but that's an impurity. You can get a little boron once in a while where chlorine should be, sometimes enough to haul out of the desert by mule team. Too many impurities, and the result will more likely be a different halide. Sedimentary rocks may seem uniform at first glance, but they are often composed of different bands of sand, silt, and clay. So, think in terms of trending to one end of the scale or the other, and less in absolutes.

## Field Studies

One big complication in identifying rocks and minerals is the tendency of the Earth's atmosphere to oxidize everything. That same oxygen we need to breathe also wreaks havoc on fresh material. Oxygen ions are always looking to hook up with another ion, preferably a metal. You've seen how a freshly sharpened knife blade starts to rust in the rain. That oxidation is also at work on cliff faces and boulders, aiding and abetting a tendency toward natural cracks, root action, freezing, and thawing. Many rocks get a reddish-brown surface in a short amount of time. First they turn color and then they fall apart.

Minerals take on distinct personalities. When you're trying to identify a mineral, you need to be able to see pure surfaces with visible crystal faces so you can measure the angles and find patterns: cubic, hexagonal, or even rhombohedral. My personal favorite is the garnet, which has a dodecahedral crystal structure, meaning it has twelve sides. You often need to break the rock and see fresh material, then look with a hand lens to see what the microcrystals can tell you.

Grand Canyon, from the North Rim
PHOTO COURTESY OF COLIN BUNN

Even then, you can't always tell what you have. There are several ways to key out a rock or crystal, such as the streak, the hardness, the luster, the color, and the weight. I've provided a glossary at the end of this book to help you with many geological terms and expressions, and I've tried to keep the text as free as possible from the really daunting vocabulary. But in many cases, I had to go with what's used by the experts.

## The Great Southwest

For the purposes of this guide, the following states make up the Southwest:

- Texas (western)
- New Mexico
- Arizona
- Utah
- Nevada
- California

Each of these states offers unique collecting opportunities, and there is a rockhounding title in the FalconGuide series for

*Sandstone buttes of Monument Valley, near the Arizona-Utah state line*
PHOTO COURTESY OF ALICE HAAR

each one. Use this guide as a companion title to help identify what you find in the field.

The topographic provinces for this region are easy enough to learn. The Great Plains that dominate the central region of North America intrude into eastern New Mexico, halted by the Rio Grande Rift that seems to be trying to cut New Mexico in half. The impressive Colorado Plateau dominates northwest New Mexico and eastern Arizona, while the Basin and Range Province begins in eastern Arizona and marches up through Southern California, parts of southern Utah, and most of Nevada. California stretches through several geologic zones, with deserts to the south, the famed Mother Lode country in the middle, and the tumultuous mélange of the western coastal region.

## Hardness

The hardness test is one of the most elementary ways to sort out samples.

| Mohs # | Material |
|--------|----------|
| 1 | Talc |
| 2 | Gypsum |
| $2\frac{1}{2}$ | *Fingernail* |
| 3 | Calcite |
| $3\frac{1}{2}$ | Copper penny |
| 4 | Fluorite |
| 5 | Apatite |
| $5\frac{1}{2}$ | *Knife blade, nail, or window glass* |
| 6 | Orthoclase feldspar |
| $6\frac{1}{2}$ | *Pyrite; steel file* |
| 7 | Quartz; agate, chert, chalcedony, and jasper; streak plate |
| $7\frac{1}{2}$ | *Garnet* |
| 8 | Topaz |
| 9 | Corundum |
| 10 | Diamond |

Try to keep a few of these common specimens on hand, and if possible, memorize the ten standard listings. I included the "non-standard" testing materials in italics.

## Be a Kind Collector

Before going any further, let's talk about safety and etiquette in the field. The biggest safety tip is to never push your car beyond its limits, and bring tools. Upgrade your tires and be religious about vehicle maintenance. Your expeditions can take you to the end of some very long and dangerous roads.

Don't treat private property like public lands; get all the permissions you need beforehand or give up and move on. Obey all signs, control your kids and dogs, pick up litter, and take only what you need when collecting. Consider bringing along an extra bucket to carry out litter you may encounter when searching road cuts.

## Additional Reference Materials

This book can help you with much of the geology you'll see in the field, because most of the visible rocks at the surface of the Earth are of common sedimentary origin. Igneous rocks are also extremely plentiful, and once you know the basic rules for those, you'll be able to screen them out as you look for nicer specimens.

*National Audubon Society Field Guide to North American Rocks and Minerals* is an 852-page treasure trove of 200+ color photographs. It's sturdy, fits in a big pocket, and takes a beating in the field. I usually leave mine at home.

For web-based research, I recommend Mindat, at mindat.org. It's a noncommercial online database of almost 39,000 mineral names, more than 300,000 photos, and 125,000 localities, and it has GPS coordinates. I rely on Google Maps, Google Earth, and the US Geological Survey.

I'm a huge fan of the Roadside Geology series because they give even the most amateur geologist a general idea about the rock exposures passing by your windshield.

*Bryce Canyon, Utah*
PHOTO COURTESY OF ROGER KASNER

These state guidebooks, published by Falcon Guides, are referenced throughout.

*Rockhounding Arizona*, by Gerry Blair, 232 pp. 2008.
*Rockhounding California*, by Gail A. Butler, 240 pp. 2012.
*Rockhounding Nevada*, William A. Kappele and Gary Warren, 264 pp. 2019.
*Rockhounding New Mexico*, by Martin Freed and Ruta Vaskys, 272 pp. 2021.
*Rockhounding Texas*, by Melinda Crow, 166 pp. 1998.
*Rockhounding Utah*, by William A. Kappele and Gary Warren, 216 pp. 2020.

I also recommend that every rockhound become familiar with geology maps, learn to use a GPS device, take great pictures, and share what you find via your favorite social networking site. See you out there!

# State Designations

It helps to know the rocks, minerals, fossils, and gems that are recognized as "official" for a state. Since it takes a lot of work and lobbying to get the state legislature to recognize something that important, you can imagine that the material is probably unique, interesting, highly sought, easy to find, or otherwise noteworthy. You wouldn't want to be in a conversation with someone from out-of-state who drove through the area and found something you didn't even know about. Here is a list for the southwestern states:

| State | Rock | Mineral | Gemstone |
|---|---|---|---|
| Texas | Petrified palmwood | Silver | Texas blue topaz |
| New Mexico | None | None | Turquoise |
| Arizona | None | Copper | Turquoise |
| Nevada | Sandstone | Silver | Black fire opal |
| Utah | Coal | Copper | Topaz |
| California | Serpentine | Gold | Benitoite |

# PART 1
# ROCKS

## Igneous Rocks: Extrusive

Use the matrix below as a starting point for understanding how igneous extrusive rocks are organized.

| Matrix of Extrusive Igneous Rocks | | | | |
|---|---|---|---|---|
| Name | Color | Grain Size | Composition | Useful Characteristics |
| Basalt | Dark | Fine or mixed | Low silica lava | Has no quartz |
| Andesite | Medium | Fine or mixed | Medium silica lava | Plagioclase with pyroxenes |
| Rhyolite | Light to medium | Fine or mixed | Very high silica | Quartz is common |
| Felsite* | Light | Fine or mixed | Medium to high silica | Rich in quartz and feldspar |
| Dacite | Light | Mixed | Medium silica | Plagioclase with hornblende |
| Pumice | Light | Fine | Sticky lava froth | Small bubbles |
| Obsidian | Dark | Fine | High silica | Glassy |
| Scoria | Dark | Fine | Runny lava froth | Large bubbles |
| Tuff | Light | Mixed | Andesite, rhyolite, or basalt | Ash fragments; sometimes welded |
| Wonderstone | Mixed | Fine | Rhyolite | Colorful, takes a polish |

*not covered in this guide

# Basalt

*Fine-grained basalt is common in the Southwest.*
PHOTO COURTESY OF THE RICE NORTHWEST MUSEUM OF ROCKS AND MINERALS

**Group:** Igneous; Extrusive
**Mineralogy:** Plagioclase feldspar, olivine, and pyroxene
**Key test(s):** Fine-grained, often with microscopic crystals only; can form dramatic columns
**Likely locale(s):** Volcanic regions

Basalt is common in many parts of the Southwest. It appears gray or light black when fresh but quickly weathers to a tan, yellow, or brown surface. It can be difficult to identify because it tends to be fine grained, so look for clues in the deposit itself. Vesicles are common; these holes sometimes fill with agate, chalcedony, or opal, in small, round "blebs" that often weather out and are collectible. Basalt can also host zeolites, quartz veins, and calcite veins under the right conditions. Andesite and rhyolite are more noted as thick, viscous lava that builds up faster, while basalt is runny and can flow hundreds of miles. While still referred to as extrusive, basalt can form in large sills and cool slowly enough to create impressive hexagons, referred to as columnar jointing. These colonnades are collectible as impressive towers known as a "Devil's Postpile" in parts of California and Nevada.

Basalt is present in every southwestern state. Utah's Black Rock Desert contains basalt flows as recent as 660 years ago. Arizona's San Francisco volcanic field, which includes Sunset Crater, contains more than 600 volcanoes and covers about 2,000 square miles. It is a major feature of the Colorado Plateau. Basalt flows are also common in the upper Grand Canyon and along the North Rim.

# Andesite

*Andesite is not as dark as basalt and has no flow lines..*
PHOTO COURTESY OF THE RICE NORTHWEST MUSEUM OF ROCKS AND MINERALS

**Group:** Igneous; Extrusive
**Mineralogy:** Intermediate; feldspars, pyroxene, and/or hornblende
**Key test(s):** Fine grained with pyroxene and plagioclase
**Likely locale(s):** Volcanic mountain ranges

Andesite is a common form of lava, usually lighter in color than basalt, and it dominates many volcanic mountain ranges, as its higher silica content causes it to pile up better, as opposed to flood basalts that flow for miles. This rock is named after the Andes Mountains, where it is a key component of those volcanoes. Andesite often exhibits a platy structure, appearing in outcrop as though it were stacked, but that isn't a telltale clue. Andesite is the extrusive equivalent of diorite, matching that rock in chemical composition. This lava is typically fine grained, with small crystals of plagioclase feldspar (such as andesine), hornblende, pyroxene (such as augite or diopside), and biotite. It makes a fine decorative rock, weathering from its light-gray appearance when fresh to a dark-gray or black color.

Andesite is common in volcanic terrain throughout the Southwest. Klondyke, Arizona, is noted for andesite dikes, and the Lake Tahoe area of northern Nevada also contains considerable andesite.

# Rhyolite

*Rhyolite is often lighter in color than other extrusive rocks.*
RHYOLITE PHOTO COURTESY OF THE RICE NORTHWEST MUSEUM OF ROCKS AND MINERALS

**Group:** Igneous; Extrusive

**Mineralogy:** Quartz, feldspar, biotite, hornblende, with accessory magnetite

**Key test(s):** Quartz crystals in fine-grained matrix

**Likely locale(s):** Common in pyroclastic flows and caldera fill

Rhyolite is another extremely common igneous rock, so learning to distinguish it from andesite and basalt is important. Its high silica content usually means rhyolite is lighter in color. In general, rhyolite is light gray, but it can appear yellow, pale yellow, and from pale red to deeper red. Rhyolite can also form in welded, compacted bands with dramatic color varieties, sometimes called wonderstone, which is carvable and takes a polish. Rhyolite cools faster than basalt and usually appears to be composed of fine-grained but distinguishable crystals. It is sometimes glassy. Flow banding is common, and crystals often show alignment under hand lens, as do any small vesicles present.

Rhyolite is common throughout the Southwest and often alternates with basalt in high-silica, low-silica bands. The Bullfrog Hills near Rhyolite, Nevada, are a classic example of gold mineralization in rhyolite host rock, requiring acid leaching to liberate the gold. Consult *Rockhounding Utah* to learn more about Topaz Mountain, where pockets in a rhyolite flow yield topaz crystals. The Birdseye "marble" from Birdseye, Utah, is actually rhyolite.

# Dacite

*Dacite is a thick, viscous lava that creates domes and dikes.*
PHOTO COURTESY OF DAVE TUCKER, WESTERN WASHINGTON UNIVERSITY

**Group:** Igneous; Extrusive
**Mineralogy:** Plagioclase feldspar, quartz, biotite, hornblende, and augite
**Key test(s):** Hornblende and plagioclase in fine-grained matrix
**Likely locale(s):** Volcanic ranges; often in dikes or sills

In composition, dacite ranks between andesite and rhyolite, typically featuring much larger crystals and a coarse, patchy appearance. Dacite is not common, so it can be difficult to get familiar with it. The color is quite variable, ranging from white, gray, and rarely black to pale red or brown, or even to deeper reds and browns. The matrix frequently appears oxidized. Mineralogically, dacite contains the same components as granodiorite, with plagioclase feldspar, quartz, biotite, hornblende, and augite. Dacite is usually found as ignimbrite or tuff, rather than in a flow, but it can display some flow banding and can also resemble obsidian.

Most volcanic regions in the Southwest, such as the San Francisco volcanic field in Arizona, contain dacite. Another notable occurrence of dacite is the Globe copper district in Arizona. Dacite is often associated with large stratovolcanoes, but it also forms as dikes and small domes.

# Pumice

*Pumice is so frothy and full of air that it floats.*

**Group:** Igneous; Extrusive
**Mineralogy:** Andesitic to rhyolitic
**Key test(s):** Floats; soft
**Likely locale(s):** Volcanic mountain ranges

Look for pumice in known volcanic regions, especially any area with recent events. Pumice is light enough to float, thanks to its frothy, glassy texture, which traps air and makes each rock buoyant. Pumice has a unique texture and feel that makes it easy to identify, and it can show dramatic flow lines. Many violent pyroclastic events spew pumice with ash and other material, causing pumice fragments to show up in breccias, tuffs, and other ash deposits. Pumice rafts are also common after volcanic eruptions in a marine environment. Collectors don't need more than a sample or two for a complete collection, unless considering industrial uses for pumice, including yard rock, potting soil, and abrasives.

Pumice is common in volcanic regions, such as the San Francisco volcanic field near Flagstaff, Arizona. Large pumice deposits are mined commercially in Madera, Inyo, Siskiyou, and Mono Counties in California; in Bernalillo and Santa Fe Counties in New Mexico; and in Coconino County, Arizona.

# Obsidian

*"Snowflake" obsidian from northern Nevada is a rare form of this volcanic glass.*

**Group:** Igneous; Extrusive
**Mineralogy:** Rhyolitic
**Key test(s):** Glassy luster, sharp edges, conchoidal fracture pattern
**Likely locale(s):** In rhyolite flows

Obsidian has no crystal structure—it chilled too fast for the atoms to align in a crystal lattice—so technically it is a glass, not a mineral. It's an easy rock to identify because it has a glassy luster when freshly chipped. When weathered or rolled in streams and rivers, however, with no fresh surface, obsidian can resemble every other rhyolite pebble on the outside. Other than the classic shiny black luster, look for evidence of flow, such as banding. The familiar fracture pattern is called "conchoidal" or shell-like; knappers have long used that tendency to break predictably to fashion amazing stone tools. The most common obsidian is jet-black, but there are other colors and forms.

According to *Rockhounding California*, at Davis Creek in Northern California, obsidian forms in long needles, and nearby Lassen Creek offers rainbow colors in many varieties. Throughout the western Basin and Range Province, fragments of obsidian formed in perlite and varnished by desert winds are known as "Apache tears," although that usage is being replaced by a more neutral "obsidianite." Refer to *Rockhounding Nevada* for vast locales where obsidian is easy to collect. Pictured here is a unique "snowflake" obsidian from Nevada. Wild Horse Canyon, the Mineral Mountains, Obsidian Hill, and Black Rock in Utah have good deposits as well.

# Scoria

*Scoria is denser and heavier than common cinder.*
PHOTO COURTESY OF THE RICE NORTHWEST MUSEUM OF ROCKS AND MINERALS

**Group:** Igneous; Extrusive
**Mineralogy:** Usually similar to basalt
**Key test(s):** Rusty, like cinders; numerous vesicles; heavier than pumice
**Likely locale(s):** Volcanic regions

Sometimes called cinder, scoria resembles pumice because both typically contain numerous air pockets, or vesicles. While pumice floats, however, scoria will sink in water. Where pumice is soft and easy to crush, scoria is usually harder and more rigid. The name comes from the Greek word for rust, and the orange-red rusty color is a common characteristic of some forms of scoria. Thus, the name "scoria" refers to a rock with a certain color or texture, rather than a specific mineralogy. The tops of lava flows, where considerable froth sometimes takes place, can harden into scoria. Similarly, some frothy ejected material from eruptions can harden as scoria. It is commonly used in landscaping and in barbecue grills.

Scoria is common in volcanic regions such as the San Francisco volcanic field in Arizona. Cinder cones near Veyo, Utah, are also noteworthy, and Nevada contains numerous examples of scoria throughout its volcanic regions, such as in Nye and Esmeralda Counties.

# Tuff

*Most of the "painted hills" in the western deserts are tuff deposits.*

**Group:** Igneous; Extrusive
**Mineralogy:** Basalt to rhyolite
**Key test(s):** Soft, unless welded; angular clasts
**Likely locale(s):** Downwind from volcanic ranges

Tuff, or lithified volcanic ash, is a common fragmental deposit around and downwind from volcanoes. Also called tephra, the term can be a bit of a catchall for the pyroclastic material ejected and deposited downwind. Tuff is usually tan to dark brown in color but can occur as pink, yellow, green, or even purple. There are varieties of tuff known as crystal tuff, containing mineral crystals; lithic tuff, with rock fragments; and welded tuff, in which pumice fragments were hot enough to compress in glassy wisps. Geologists also denote rhyolitic tuff, andesitic tuff, basaltic tuff, ultramafic tuff, trachyte tuff, and others. Tuff is usually found as a breccia, with a fine- to medium-grained matrix, and can contain large angular material such as pumice, volcanic bombs, and other debris. Tuff beds can be prime fossil-hunting locales, containing petrified wood, leaves, bones, and other organic debris.

Many of the "painted hills" in the western United States are tuffs and volcanic ash, including the Painted Desert area near Hoover Dam in Mohave County, Arizona. The major volcanic regions of the Southwest each contain significant tuff deposits.

# Wonderstone

*"Wonderstone" is a term for rhyolitic ash deposits with flow patterns.*

**Group:** Igneous; Extrusive

**Mineralogy:** Quartz, feldspar, biotite, hornblende, with accessory magnetite

**Key test(s):** Scratch test for best specimens

**Likely locale(s):** Common in pyroclastic flows and caldera fill

Wonderstone is a form of rhyolite that features distinctive flow patterns, usually red, white, pink, yellow, or tan. Rather than a lava flow, wonderstone is a form of volcanic ash that is called a tuff. In this case, the tuff particles are said to be welded or cemented. It forms during explosive volcanic eruptions when hot, liquid material is ejected and sticks together upon landing. If such rocks are buried under more hot material, there is a good chance the deposit will form into a solid glassy deposit, and the higher the hardness, the better the material is for lapidary. Wonderstone is carvable and sometimes takes a polish, but typically the material is soft and porous and notable only for the distinctive flow banding.

Wonderstone is found in abundance at Wonderstone Mountain near Fallon, Nevada, and at Vernon Hills, Utah, and in places in northern California. *Rockhounding Nevada* and *Rockhounding Utah* both contain collecting information.

# IGNEOUS ROCKS: INTRUSIVE

Below is a table comparing the various igneous intrusive rocks.

| Matrix of Intrusive Igneous Rocks | | | | |
|---|---|---|---|---|
| Name | Color | Grain Size | Composition | Features |
| Granite | Light, often pink | Coarse | Feldspar, quartz, mica, and hornblende | Wide range of color; coarse grain size |
| Syenite* | Light | Coarse | Mostly feldspar and minor mica | Like granite but no quartz |
| Tonalite* | Light to medium; salt-and-pepper | Coarse | Plagioclase feldspar and quartz plus dark minerals | Limited alkali feldspar |
| Porphyry | Any | Mixed | Large grains of feldspar, quartz, olivine, pyroxene | Large grains in a fine-grained matrix |
| Gabbro | Medium to dark | Coarse | High calcium | No quartz; limited olivine |
| Diorite | Medium to dark | Coarse | Low calcium | Limited quartz |
| Peridotite | Dark; often greenish | Coarse | Olivine present | Dense; 40+% olivine |
| Pyroxenite* | Dark | Coarse | Pyroxene | Rich in pyroxene |
| Dunite* | Green | Coarse | Olivine dominant | Dense; 90+% olivine |
| Kimberlite* | Dark | Coarse | Source of diamonds | Found in "pipes" |
| Pegmatite | Any | Very coarse | Usually granitic | Dikes; small intrusions |

*not covered in this guide

# Granite

*Pink granite gets its color from excess potassium feldspar.*
PHOTO COURTESY OF THE RICE NORTHWEST MUSEUM OF ROCKS AND MINERALS

**Group:** Igneous; Intrusive
**Mineralogy:** Quartz, feldspar, hornblende, biotite, and magnetite
**Key test(s):** Often pinkish; exfoliation
**Likely locale(s):** Mountainous terrain

Granite is one of the most common and recognizable igneous intrusive rocks. It is usually coarse grained and is primarily composed of feldspar, hornblende, and quartz. It can be massive or display zoning, depending on how close it came to the exterior of the intrusion. It is sometimes a salt-and-pepper rock; some have a little more feldspar here, while others have a lot of dark hornblende there. Granite may occur as minor intrusions, as larger plutons, or as huge batholiths taking up tens of thousands of square miles. It can also form impressive cliffs, such as Yosemite's El Capitan. One field clue is the tendency of granite to weather by exfoliation, where thin sheets peel off like the skin of an onion. Granite isn't highly collectible, but it is easily slabbed into tabletops and counters or used for buildings and monuments because granite takes a high polish. Probably the most interesting aspect of granite is its association with economic ore deposits, often found along the margins of granite intrusions, and the presence of pegmatites.

California's Yosemite National Park is one notable granite deposit in the Southwest; other sizable granite outcrops include Arizona's Granite Mountain, near Prescott; the Sandia Mountains in New Mexico; the Granite Mountains near Gerlach, Nevada; and Granite Mountain in the Wasatch Range of Utah. Many of Arizona's famed copper deposits are associated with porphyritic granitic rocks.

# Porphyry

*Andesite porphyry marked by presence of large crystals.*
PHOTO COURTESY OF THE RICE NORTHWEST MUSEUM OF ROCKS AND MINERALS

**Group:** Igneous; Intrusive
**Mineralogy:** Typically basalt or rhyolite
**Key test(s):** Contains large quartz or feldspar crystals
**Likely locale(s):** Volcanic regions

A rock is said to be a porphyry when it appears to contain notable large crystals in a fine-grained matrix. The chemical composition of the large crystals and the matrix may be similar, but something occurred to get two different phases of cooling in these rocks. The large crystals are usually 2 mm or greater in size and are typically distinguishable as either quartz or feldspar. This apparently represents a time when the liquid magma cooled slowly, deep inside the Earth, because large crystals typically take a long time to form. But subsequent events led to the still-liquid magma rising rapidly to the surface and then cooling quickly, so that the large crystals were frozen in place. Most igneous rocks seen in the field show this texture if studied closely, but striking porphyries will catch your eye. This texture is seen in many volcanic rocks throughout the Southwest.

# Gabbro

*Gabbro is usually dark and very dense.*
PHOTO COURTESY OF THE RICE NORTHWEST MUSEUM OF ROCKS AND MINERALS

**Group:** Igneous; Intrusive
**Mineralogy:** Feldspar, hornblende, biotite, and magnetite
**Key test(s):** Dark with coarse grains
**Likely locale(s):** Mountainous terrain

Gabbro is an interesting intrusive rock, distinguished by very coarse texture and a range of dark colors. One helpful characteristic is the presence of green olivine, which contrasts with the usual black or dark-gray color. Gabbros can be dark red, however, so that color test isn't precise. Gabbro sometimes makes up the base unit of massive, layered intrusions such as the famed Stillwater Complex in Montana or the Skaergaard intrusion of east Greenland. Perhaps the most famous layered intrusive with gabbro at the base is the storied Bushveld Complex of South Africa, the source of much of the world's platinum. These complex intrusions are often rich in chrome as well. Layering is sometimes visible in fresh exposures, but gabbro tends to weather faster and that can also aid in field identification. Other names include diabase, dolerite, and black granite. Gabbro takes a polish and can be fashioned into dark countertops, monuments, and statues.

The Jurassic gabbroic complex of Churchill and Pershing Counties in Nevada is noteworthy for this rock. The eastern Mojave Desert contains Precambrian gabbros, as do the San Gabriel Mountains.

# Diorite

*Common diorite is easy to confuse with granite but is typically darker and fine grained.*
PHOTO COURTESY OF THE RICE NORTHWEST MUSEUM OF ROCKS AND MINERALS

**Group:** Igneous; Intrusive

**Mineralogy:** Feldspar, hornblende, biotite, and pyroxene

**Key test(s):** Dark smears, unlike granite; often in dikes and sills

**Likely locale(s):** Mountainous terrain

Diorite is a medium- to coarse-grained igneous intrusive rock, made up of common minerals such as plagioclase feldspar and pyroxene. It is usually medium grained, so crystals are recognizable under hand lens, but it can be coarse, with zoning common. Diorite is usually gray to dark gray, but it can be lighter, and a green or brown tint isn't out of the question. Salt-and-pepper colors are very common. Compared to granite and granodiorite, plain diorite is not common, and it is often found as dikes, sills, and stocks at the margins of large granite batholiths. Because diorite is relatively hard, it takes a good polish and was used by ancients for inscriptions, such as for the Code of Hammurabi, which was carved into black diorite, and the Rosetta Stone, which was carved into a granodiorite.

Arizona's Big Horn Mountains contain significant diorite, but just about every major area in the Southwest that contains granite also contains diorite.

# Peridotite

Chunky peridotite showing green olivine crystals.
PHOTO COURTESY OF THE RICE NORTHWEST MUSEUM OF ROCKS AND MINERALS

**Group:** Igneous; Intrusive
**Mineralogy:** Olivine and pyroxene
**Key test(s):** Green; coarse
**Likely locale(s):** Ancient terrains

Peridotite is a dark, dense rock made up of olivine and little else. Peridot is the gem variety of olivine and lends its name to this rock. Peridotite can be as much as 90 percent olivine; higher concentrations would infer a dunite. The olivine is usually present as large crystals; thus a coarse texture is another key distinguishing characteristic. It can form in layers or as massive structures. Peridotite is one of the most exotic rocks available in the field, as it is closely associated with the Earth's interior and is believed to be the primary rock found in the upper mantle. Since olivine reacts quickly with water and oxygen, peridotite is not very stable, and upon exposure to the atmosphere, it quickly starts converting to serpentinite.

Peridotite is known throughout the Southwest region. Kilbourne Hole, New Mexico, is said to be a classic peridotite dredged up from the mantle. Other locales include the southern California Coast Range; Park City and Moses Rock, Utah; the eastern Sierra Nevada; and San Carlos, Arizona.

# Pegmatite

*Pegmatite is noted for large crystals of feldspar, mica, schorl, and other minerals.*
PHOTO COURTESY OF THE RICE NORTHWEST MUSEUM OF ROCKS AND MINERALS

**Group:** Igneous; Intrusive
**Mineralogy:** Quartz, feldspars, altered olivine, rare garnets, orthopyroxene, and chrome diopside
**Key test(s):** Rare garnets; very coarse crystals
**Likely locale(s):** Older continental crust

Pegmatites sometimes host the most prized of gems, such as rubies, emeralds, and sapphires, so they are important to learn about because of their potential reward. The word refers to their coarse texture with large crystals, sometimes resembling a patchwork quilt. Pegmatites are usually found at the margins of large granite bodies and contain the same minerals, such as alkali feldspar and quartz. There are varieties, such as granite pegmatite and nepheline syenite pegmatite, but pegmatites are all usually white due to the amount of feldspar present. Pegmatites can also appear light yellow, tan, or even gray. Zoning is common, with vugs and cavities highly sought.

Famous pegmatite belts and outcrops in the Southwest start with the Himalaya Mine in Southern California, covered in *Rockhounding California*. The Harding Mine in Taos County, New Mexico, is a famous pegmatite deposit, also covered in that state's guidebook. Mohave, Maricopa, and Yavapai Counties in Arizona all have significant pegmatites. Honeycomb Hills in western Utah is noteworthy, and Nevada also has numerous pegmatites.

# METAMORPHIC ROCKS

Use the matrix below as a starting point for understanding how metamorphic rocks are organized.

| Name | Hardness | Foliation | Grain Size | Color | Other |
|------|----------|-----------|------------|-------|-------|
| Soapstone* | Very soft | Foliated | Fine | Light | Greasy |
| Slate | Soft | Foliated | Fine | Dark | Striking sound |
| Phyllite | Soft | Foliated | Fine | Dark | Shiny and crinkly |
| Serpentinite | Soft | Nonfoliated | Fine | Green | Shiny, mottled |
| Marble | Medium | Nonfoliated | Fine | Light | Calcite or dolomite by acid test |
| Greenstone | Medium | Relicts | Fine | Light | Common |
| Argillite | Hard | Relicts | Fine | Mixed | Common |
| Mylonite* | Hard | Foliated | Coarse | Mixed | Crushed and deformed |
| Schist | Hard | Foliated | Coarse | Mixed | Large, deformed crystals |
| Gneiss | Hard | Foliated | Coarse | Mixed | Banded |
| Migmatite* | Hard | Foliated | Coarse | Mixed | Melted |
| Amphibolite | Hard | Foliated | Coarse | Dark | Hornblende |
| Hornfels | Hard | Nonfoliated | Fine or coarse | Dark | Dull, opaque |
| Eclogite* | Hard | Nonfoliated | Coarse | Red and green | Dense; garnet and pyroxene |
| Quartzite | Hard | Nonfoliated | Mixed | Light | Quartz—no fizzing |

*not covered in this guide*

# Quartzite

*Typical specimen from Quartzsite, Arizona*
PHOTO COURTESY OF THE RICE NORTHWEST MUSEUM OF ROCKS AND MINERALS

**Group:** Metamorphic; regional
**Mineralogy:** Muscovite mica, chlorite; pyrite is a frequent accessory
**Key test(s):** Foliation; visible crystals; density
**Likely locale(s):** Metamorphic terrain

Quartzite is a common pebble in many streams and rivers because it is hard and strong. Quartzite is characterized by a vitreous luster and is usually found as a round, white rock, especially if pure. Other forms of quartzite can be light gray, and it can even run to pink or light brown. Quartzite is usually fine grained, but sometimes crystals are actually visible and the rock can be medium grained. Quartzite deposits are typically thick, massive units, resistant to erosion and forming prominent bluffs. Quartzite will break across bedding planes, not along them. The hardness is typical of quartz, at 7, but it can be softer if significant amounts of feldspar and calcite are present. This rock takes a nice polish and has some industrial applications as landscape rock, tiles, flooring, and even ballast. Some collectors confuse quartzite pebbles for agate because they share similarities, but reserve the term "agate" for translucent, banded quartz.

This rock is found throughout the Southwest, such as in Utah's Sessions Mountains, but the most famous locale is in the hills near Quartzsite, Arizona, home to the popular annual rock-and-gem collectors' gathering. Consult *Rockhounding Arizona* for more information.

# Marble

*Bright white marble takes a nice polish.*

**Group:** Metamorphic; regional
**Mineralogy:** Muscovite mica, chlorite; pyrite is a frequent accessory
**Key test(s):** Foliation; visible crystals; density
**Likely locale(s):** Metamorphic terrain

There are dozens of varieties of marble, but in general it is simply a limestone or dolomite that has been highly metamorphosed, usually into a fine- to medium-grained specimen with a soft, vitreous luster. Marble is normally white, sometimes brilliantly so, but it is often darker thanks to black, green, red, pink, or even yellow staining from various impurities. It often appears as though it is shot through with veins of darker material that originated in the limestone bed as clay, sand, silt, or chert. This produces a lined or "marbled" look. Because limestone deposits are often quite large, it's logical that marble is usually massive, with foliation, banding, and streaking all common. Marble has a hardness of only about 3 but can range from 2 to 5. Calcite marble effervesces with cold hydrochloric acid. Dolomitic marble must be pulverized, and the acid must be heated to witness the acid effect.

Marble was once mined near Prescott and east of Wilcox in Arizona. The Carrara marble locale in California is another good source, as are deposits near Newhouse and Provo City, Utah. Many areas of California have significant marble deposits, including San Bernardino, Tuolumne, Sonoma, Riverside, Amador, and Inyo Counties. *Rockhounding Utah*, *Rockhounding Nevada*, and *Rockhounding California* have more information.

# Serpentinite

*Serpentinite is a mixture of several related minerals. This specimen is from Serpentine Canyon in Northern California.*

**Group:** Metamorphic; regional
**Mineralogy:** Muscovite mica, chlorite; pyrite is a frequent accessory
**Key test(s):** Green, greasy, streaky, and soft
**Likely locale(s):** Metamorphic terrain

Serpentinite is a collective term for about twenty different magnesium iron phyllosilicates commonly found in varying percentages in most deposits. Serpentine is the most well-known constituent, but other common minerals include antigorite, lizardite, and chrysotile. Asbestos is a common constituent, so use caution. This rock is usually green, with streaks of black, yellow, white, or gray. It is very dense, which distinguishes it from soapstone. The best field test is serpentinite's slick, greasy feel, which is a key characteristic. The structure is usually massive, if jumbled, and serpentinite is famous for "slickensides" where micro faulting and zones of movement result in polished, "slick" surfaces. The hardness for serpentinite is usually 4 or less on the Mohs scale, sometimes as low as $2\frac{1}{2}$, depending on how much hard quartz is present. It is carvable, and its varied hues make for interesting sculptures, even if it is harder than soapstone.

California has significant serpentinite zones, sometimes associated with chromite, cinnabar, and magnesite. Serpentine is the state rock of California, and it is common in the state's Coast Range. In Arizona, Gila County and Cochise County are both noted for varieties of serpentinite. Utah, Nevada, and New Mexico all contain serpentinite exposures. Consult the relevant state guidebook for more information.

# SEDIMENTARY ROCKS

Use the matrix below as a starting point for understanding how sedimentary rocks are organized.

| Name | Hardness | Grain Size | Composition | Features |
|---|---|---|---|---|
| Coal | Soft | Fine | Carbon | Black; burns |
| Shale | Soft | Fine | Clay minerals | Splits in layers |
| Limestone | Soft | Fine | Calcite | Fizzes |
| Dolomite | Soft | Coarse or fine | Dolomite | Fizzes if powdered |
| Coquina* | Soft | Coarse | Fossil shells | Bits and pieces |
| Graywacke/ wacke | Hard or soft | Mixed | Mixed sediments with rock grains and clay | Gray or dark and dirty |
| Concretion | Hard, round | Fine | Lime-rich matrix; organic material inside | Round "cannonballs" |
| Conglomerate | Hard or soft | Mixed | Mixed rock and sediment | Rounded rocks glued together |
| Breccia | Hard or soft | Mixed | Mixed rocks and sediment | Sharp-edged conglomerate |
| Sandstone | Hard | Fine-medium | Clean quartz | Grainy |
| Arkose* | Hard | Coarse | Quartz and feldspar | Quartzy sandstone |
| Mudstone | Soft | Fine | Clay-rich hardened mud | Hard to see particles |
| Siltstone | Hard | Fine | Very fine sand; no clay | Gritty |
| Chert | Hard | Fine | Chalcedony | No acid fizz |

*not covered in detail in this guide*

# Coal

*Common coal has a dark, shiny appearance; the lower the grade, the messier it is.*

**Group:** Sedimentary
**Mineralogy:** Carbon
**Key test(s):** Oily smell; dusty; light heft
**Likely locale(s):** Metamorphic terrain

Coal actually spans the boundary between a sedimentary rock and a metamorphic rock. Even at high grades, coal is very soft; its overall hardness ranges from 1 to $2\frac{1}{2}$. Once coal bakes into the anthracite stage, it is less greasy, but coal dust is always a problem due to its tendency to be brittle and friable. Further metamorphism leads to graphite and pure carbon, but diamonds exist in kimberlite, not from metamorphosed coal. The higher the quality, the fewer relicts are present, and color can get more interesting, with purple iridescence. High-grade coal is up to 98 percent hydrocarbons, but sulfur and nitrogen are almost always present.

The San Juan and Raton Basins of New Mexico are significant producers of coal; Arizona's Black Mesa coal deposit is also important. Central Utah contains large coal deposits. California coal mining has never shown much promise other than short-lived mines near Mount Diablo and Tesla.

# Shale

*Common shale often has bedding planes and ripple marks.*

**Group:** Sedimentary
**Mineralogy:** Muscovite mica, chlorite; pyrite is a frequent accessory
**Key test(s):** Foliation; visible crystals; density
**Likely locale(s):** Sedimentary basins

Increasing metamorphism ⟶

| Mudstone | Shale | Slate | Phyllite | Schist | Gneiss |
|----------|-------|-------|----------|--------|--------|

Shale is still classified as a sedimentary rock, but it has started to harden in distinguishable ways. Varieties include oil shale, calcareous shale, carbonaceous shale, and others. Shale is usually dark but not always, as its color depends on the source rock from which it hardened. Shale is noted for the way it fractures into plates, along bedding planes, where evidence of past water action such as waves, ripples, cracks, and footprints are all observable. Shale beds are easy to spot, although they can be confused with platy andesite. Look for evidence of fossils or water features for one differentiator. Note that some rockhounds use the slang "shale" for unwanted rock, no matter what its true composition.

The Mancos Shale dominates in New Mexico's San Juan Basin and nearby Arizona. The Hermit Shale occurs in northern Arizona. Utah hosts interesting shale deposits in Uintah and Grand Counties, and the Green River Formation in Utah has thick shale sequences. The Chainman Shale and White Pine Shale of Nevada are well known.

# Limestone

*Fossiliferous limestone*

**Group:** Sedimentary
**Mineralogy:** Chemical deposition
**Key test(s):** Fizz test; fossils
**Likely locale(s):** Old sea basins

Limestone is one of the most common sedimentary rocks, and it comes in many varieties. Chalk is fine grained, derived from the skeletons of tiny sea creatures, while coquina contains large, abundant shell fragments. Travertine is typically banded and colorful, while oolitic limestone refers to tiny orbs, or oolites, which are very small concretions. The term "marl" is used to describe a limestone with a high percentage of silicates. Limestone is often marked in the field by pitted and pockmarked outcrops. The hardness for limestone is in the range of 3 to 4. The most reliable test is fizzing in cold, diluted hydrochloric acid. Typically, limestones are light gray, but if iron is present, they can stain reddish, trending to darker gray. Limestone is usually dense, and bedding can be obvious.

Limestone is common in the Southwest, and there are numerous fossil-hunting locales noted in the guidebooks. The Permian-aged Capitan Reef in New Mexico hosts the famed Carlsbad Caverns. The Kaibab Formation stretches across northern Arizona, southern Utah, east-central Nevada, and southeastern California. Lime from San Bernardino, limestone and dolomite from southwestern Nevada, and marl from near Pyramid Lake in Nevada all have been significant.

# Dolomite

*Common dolomite rarely has fossils and contains much more magnesium.*

**Group:** Sedimentary
**Mineralogy:** Chemical
**Key test(s):** Fizz test
**Likely locale(s):** Common

Dolomite, sometimes called dolostone, is typically tan or light gray, but varieties can range all the way to pink and dark gray. It is typically dense like limestone, its cousin, but there is usually less evidence of grains or fossils. Typically, dolomite has a very fine texture because it doesn't have the shell fragments or oolites of limestone. Limestone and dolomite share many characteristics, but the big difference is that dolomite has substituted more magnesium for calcium, via circulating groundwater. It can appear as bands and grade slowly into limestone. The best evidence for separating dolomite from limestone is with hydrochloric acid: Dolomite will only fizz if the acid is hot and the material has been ground into a powder, while limestone is much more reactive. Dolomite is used for fertilizers high in calcium and magnesium and as an easy source for magnesium.

Dolomite is common throughout the Southwest, wherever limestone beds exist. The Carlsbad Caverns area of New Mexico contains dolomite occurrences, while Utah, Nevada, and Arizona also host numerous deposits.

# Graywacke

*This common graywacke from northern Nevada resembles a dirty sandstone.*

**Group:** Sedimentary
**Mineralogy:** Clastic
**Key test(s):** Clay oxidation
**Likely locale(s):** Sedimentary basins

Graywacke, a specific type of wacke (pronounced WACKY), is a sedimentary rock composed of poorly sorted, rounded, or broken bits of shale, slate, basalt, granite, or chert in a fine clay matrix. It is usually gray, hence the name, but can be dark gray or reddish with the presence of iron. There is typically no sorting in a graywacke, giving it an unusual appearance among conglomerates and sandstones. Quartz and feldspar grains are common, but the grain size and structure of graywacke varies quite a bit. Think of it as a dirty sandstone, which typically dates to the Silurian period in Europe.

Exposures in the Southwest are known but not common. Humboldt, Washoe, and Pershing Counties in Nevada contain graywacke exposures. Another locale is at Baker Beach near the Presidio in California.

# Concretion

*Concretions are usually empty but sometimes contain marvelous treasures.*
PHOTO COURTESY OF THE RICE NORTHWEST MUSEUM OF ROCKS AND MINERALS

**Group:** Sedimentary
**Mineralogy:** Clastic environment but chemical too
**Key test(s):** Round, heavy, splittable
**Likely locale(s):** Sandstone

Concretions are the lottery ticket of the fossil world. They are typically round or oval structures that can range from tiny, pea-size orbs to large, beach ball–size or larger monsters. A concretion typically forms when some organic debris starts rolling around in a lime-rich mud within an active marine bay or lagoon. The mud will stick to organic material, and soon the original is completely coated. Further agitation results in a large, round object, which often winds up in a sandstone or mudstone, sometimes in striking zones. Fossil hunters value them because if there is a fossil snail, crab, or even whale skull inside, it is typically well preserved. Search for concretions in known marine sandstones, siltstones, and mudstones—especially where the term "cannonball" is used to name local geography.

Known concretion areas in the Southwest include Bowling Ball Beach in Mendocino County, California, Fossil Insect Canyon in the Barstow Formation of the Mojave Desert, the Pumpkin Patch near Anza-Borrego Desert State Park, and near El Centro in the Imperial Valley. Also consider the Frontier Formation in northwest Utah and Nevada. Southern Utah's Navajo Sandstone has unique iron concretions across a wide area. The famed Moqui Marbles in Utah's Spencer Flat are small iron concretions.

# Conglomerate

*Typical conglomerate, with rounded clasts cemented in matrix*
PHOTO COURTESY OF THE RICE NORTHWEST MUSEUM OF ROCKS AND MINERALS

**Group:** Sedimentary
**Mineralogy:** Large clasts in cement matrix
**Key test(s):** Rounded pebbles
**Likely locale(s):** Nonmarine

Conglomerates are typically found as hard, haphazardly sorted beds of large, rounded pebbles glued together in a fine- to medium-grained matrix. If the pebbles are still angular and edgy, the rock would be called a breccia. This is a clastic rock, and the clasts themselves can be big or small. By definition, the rocks and pebbles are greater than 0.08 inch in diameter, but they can range to large cobbles and boulders. Because conglomerates are typically deposited in nonmarine waters that are very active, the clasts are usually hard material such as quartzite, chert, or flint, which can survive that much churning. The matrix of a conglomerate is often dominated by silica or calcite, making this rock extremely tough and resistant to erosion.

Conglomerates are common across the Southwest but rarely significant enough to lend their name to an entire formation. One exception is the Pala Conglomerate in Southern California. Others include the Gila Conglomerate, the Beavertail Conglomerate, and the Glance Conglomerate, all found in Arizona and New Mexico.

# Breccia

*Breccia is composed of angular clasts that have not had their corners rounded off.*

**Group:** Sedimentary
**Mineralogy:** Large clasts in cement matrix
**Key test(s):** Rounded pebbles
**Likely locale(s):** Nonmarine

Unlike conglomerates, breccias consist of broken rock pieces that are still angular and sharp edged. This is usually evidence that the source material is located nearby, as significant rounding hasn't started. These rocks are typically related to tuff and volcanic ash and are usually lighter in color. They can be made up of mixed material, but the general rule is that the rock fragments are coarse, angular, and varied. Breccias are usually found in thick, massive beds, possibly interbedded with fine-grained tuff. Breccias are usually seen as evidence of explosive forces, such as meteorite-impact breccias, and from volcanic explosions. Breccias can also depict evidence of significant movement along faults.

Sedimentary breccias are common across Arizona, while New Mexico breccias are mostly volcanic in nature. The Alamo impact breccia in Devonian rocks of the Pahranagat Range of southeastern Nevada is a famous site for meteorite hunters.

# Sandstone

*Valley of Fire State Park in Nevada contains vivid red sandstone landforms.*

**Group:** Sedimentary; clastic
**Mineralogy:** Large quartz or feldspar clasts in cement matrix
**Key test(s):** Rounded pebbles
**Likely locale(s):** Nonmarine

Sandstone is one of the most common sedimentary rocks, and it is found throughout the Southwest. Frequently, sandstones in the field are found interbedded with mudstone, limestone, shale, and other sedimentary rocks. By definition, all sandstones are marked by a grain size of 0.05 to 2.0 mm, which makes it medium grained compared to mudstone and siltstone but not super-coarse like a conglomerate. Sandstone is quite varied in color and found in hues of gray, brown, red, yellow, and even white. Sandstone bluffs and beds usually show at least some evidence of sorting. What holds the clasts or grains together in a sandstone is usually silica, but calcite or even iron oxide will also serve. Sandstone is relatively easy to shape and fashion into buildings and walls, and it is often used as a decorative stone. Fossil hunters inspect sandstone for leaves, shells, and bones.

The Wingate, Moenkopi, and Navajo Sandstones, to name a few, are familiar in Arizona; New Mexico has numerous sandstone formations as well. Upper Cretaceous sandstone is common in the Upper Great Valley Sequence of California. Utah and Nevada also host numerous sandstones of various ages. Valley of Fire State Park in Nevada consists of bright red Aztec sandstone outcrops.

# Mudstone

*Fossiliferous mudstone*

**Group:** Sedimentary
**Mineralogy:** Mud, silt, sand
**Key test(s):** Very fine grained
**Likely locale(s):** Seashore

Mudstone is a common sedimentary rock, usually light tan or light gray, but it can be darker or redder. It is usually dense and sometimes has organic material present in its extremely fine-grained matrix. It is composed of tiny particles of silt or clay, hence the name. When dry, mudstone can appear massive, but when wet the bedding planes are easier to pick out. Mudstones are prone to weather quickly, and their cliffs recede markedly in wet, marine climates. Fossils tend to be well preserved in mudstones but often must be immediately prepped in the field with preservative or they are immediately attacked by moist air and reduced to chalk.

Mudstone of all ages, dating back to the Precambrian era, is common across the Southwest.

# Siltstone

*Siltstone with fossil trilobite, Delta, Utah*
PHOTO COURTESY OF THE RICE NORTHWEST MUSEUM OF ROCKS AND MINERALS

**Group:** Sedimentary; clastic
**Mineralogy:** Very fine
**Key test(s):** Visible bedding planes
**Likely locale(s):** Marine

Siltstone is a common sedimentary rock that ranges from light gray and light brown in color, but it can be darker if more organic material or iron staining is present. It is usually dense, and under a microscope contains small particles of silt or clay. Bedding planes are often easy to see, especially if wet. Look for clay, such as kaolinite, feldspars, and quartz grains, plus mica flakes, but at a very small scale. Siltstone cliffs recede quickly in wet climates. Fossils tend to be poorly preserved in these rocks and often must be immediately prepped in the field with a polyvinyl acetate coating.

Siltstone is common across the Southwest, grading into sandstone, mudstone, and coarser rock units.

# Chert

*Chert from northern Nevada*

**Group:** Sedimentary
**Mineralogy:** Silica
**Key test(s):** Hardness 7; organic nature; ooze origins
**Likely locale(s):** Nodules in limestone

Like most quartz incarnations, chert has a hardness of 7. It is commonly white in color, but banded chert can take on reds, yellows, and even greens. Chert is a by-product of organic ooze from the seafloor hardening into a rock, while jasper results from circulating silica solutions, usually in basalt. At high magnification some cherts display tiny little skeletons of microorganisms. When black, chert is more commonly referred to as flint and was easily fashioned into tools and weapons by Stone Age artisans. Chert is dense, and smooth when polished, but can feel very rough where exposed. Bedded or ribbon cherts in basalts and sandstones are prized by collectors, who look for seams of material big enough to slice with a rock saw and polish into cabochons.

The Lake Range quarry in Washoe County, Nevada, is a famed chert locale with archaeological significance, but there are numerous others throughout that state, and they are frequently associated with economic ore deposits. Chert is common with the limestone at Carlsbad Caverns, New Mexico. Utah, Arizona, and Southern California all host chert-collecting locales as well. Ribbon chert is common in the Eel River and covered in *Rockhounding California*.

# PART 2

# MINERALS

## Actinolite

*Dark green actinolite crystals often form a spray.*

$Ca_2(Mg,Fe)_5Si_8O_{22}(OH)_2$
**Family:** Amphibole silicates
**Mohs:** 5–6
**Specific gravity:** ~3
**Key test(s):** Green, fibrous, brittle
**Likely locale(s):** Metamorphic terrain

Actinolite is usually green, ranging from pale green to dark green, sometimes with yellow tinting. It can display as long, matted blades, usually quite brittle, and the blades can be parallel or create spectacular sprays. Like so many other minerals, actinolite occurs in a series, with one end being rich in magnesium and the other rich in iron, and near-infinite percentages between the two extremes. Actinolite sits in the middle; magnesium-rich actinolite is called tremolite, while iron-rich actinolite gets the unimaginative name ferro-actinolite. These rocks are always associated with asbestos, so beware. Related rocks include nephrite jade.

Actinolite is common across the southwestern states.

# Agate

*Red plume agate from Brewster County, Texas*
PHOTO COURTESY OF THE RICE NORTHWEST MUSEUM OF ROCKS AND MINERALS

**Quartz, SiO$_2$**
**Family:** Cryptocrystalline quartz
**Mohs:** $6\frac{1}{2}$–7
**Specific gravity:** 2.65
**Key test(s):** Banding; translucence
**Likely locale(s):** Veins and blebs in basalt

Agate is a term loosely used by rockhounds to describe any clear or translucent quartz, but the term is interchangeable with chalcedony, and chalcedony is the preferred term among mineralogists. However, agate hunting, agate picking, and agate digging are such a part of our rockhounding lexicon that the term will never go away. Agate is one of the most popular and collectible forms of quartz, and it comes in a wide variety, including lace agate, moss agate, banded agate, and others. Usually the term "agate" should be reserved for translucent, banded quartz, and usually the banding is quite distinctive. Typical colors are clear, red, yellow, or a pleasing, highly sought light blue. True agate has a fine grain and excellent color.

Countless areas of the Southwest contain prized agate—collecting locales are included in each state's guidebook. "Patriotic agate" is a noteworthy variant covered in *Rockhounding Arizona*.

# Amethyst

*Purple amethyst showing classic quartz crystal structure*

**Quartz, SiO$_2$**
**Family:** Crystalline quartz
**Mohs:** 7
**Specific gravity:** 2.65
**Key test(s):** Purple; harder than fluorite
**Likely locale(s):** Quartz veins

Amethyst is a form of crystalline quartz, like smoky quartz, only amethyst is typically purple, and sometimes vividly violet. Iron impurities impart the purple color. Like common quartz, amethyst reaches a hardness of 7 on the Mohs scale. Crystals are hexagonal and often sharply terminated

Much of the world's finest amethyst comes from large, amethyst-lined geodes in Brazil, but there are numerous amethyst locales covered in the guidebooks for the Southwest states. In west Texas the Woodward Ranch and Needle Peak host amethyst deposits. The Black Hawk Mine in New Mexico is probably that state's best deposit to explore. In Utah, Agate Switch and the San Rafael District have received attention. Southern Arizona has numerous locales; Arizona's Four Peaks Amethyst Mine offers tours and fee-based collecting. Nevada has locales near Getchell and Reno, while Southern California hosts deposits in the Kingston Range, in Escondido Canyon, and in the Diablo Range.

# Apatite

*Well-formed apatite crystal*
PHOTO COURTESY OF THE RICE NORTHWEST MUSEUM OF ROCKS AND MINERALS

**Calcium phosphate, $Ca_5(PO_4)_3(F,Cl,OH)$**
**Family:** Phosphates
**Mohs:** 5
**Specific gravity:** 3.1–3.2
**Key test(s):** Hardness test
**Likely locale(s):** Pegmatites

Apatite is the name given to a group of phosphate minerals such as fluorapatite and chlorapatite, so the chemical formula can vary according to how much of one element or another is present. As expected, the color varies. Darker green, purple, or violet are common, as are red, yellow, and pink. Apatite is a common mineral in igneous rocks; larger crystals occur in pegmatites. Fluorite has similar coloration but isn't as hard, and the crystal habit is cubic, while quartz is harder. Interestingly, apatite is the hardening agent for bones and teeth.

Apatite minerals are known throughout the Southwest, from Big Bend Country in Texas to San Bernardino County in California. Apatite is frequently a secondary mineral in economic ore deposits, hence its widespread distribution. Noted locales include the Himalaya Mine in Southern California and at Iron Mountain in Utah.

# Aragonite

*Aragonite sometimes displays attractive branching.*
PHOTO COURTESY OF THE RICE NORTHWEST MUSEUM OF ROCKS AND MINERALS

**Calcium carbonate, $CaCO_3$**
**Family:** Carbonates
**Mohs:** $3\frac{1}{2}$–4
**Specific gravity:** 2.95
**Key test(s):** Fibrous appearance
**Likely locale(s):** Fossil shells, stalactites

Aragonite is another form of calcium carbonate, but unlike its related cousin calcite, aragonite usually occurs thanks to biological processes. Crystals are usually white or milky white, but with impurities, a vast range of colors are possible. Some mollusks form their shells completely with aragonite; other mollusks alternate between calcite and aragonite. Aragonite is not strictly biological, however. For example, the stalactites of Carlsbad Caverns are aragonite. The flowering pattern is often distinguishing. Interestingly, aragonite is not very stable, in a geologic sense—there are no deposits of aragonite known that are older than the Carboniferous period, which ended about 300 million years ago.

Aragonite is common around the limestone deposits of the Southwest, and particularly at Carlsbad Caverns in New Mexico, where it forms stalactites, or "cave flowers."

# Augite

*Large augite crystals are rare, but augite is a common component of igneous rocks.*

**Silicate, Ca,Na(Mg,Fe,Al)(Al,Si)$_2$O$_6$**
**Family:** Silicates
**Mohs:** 5–6
**Specific gravity:** 3.2–3.6
**Key test(s):** Black, stubby crystals
**Likely locale(s):** Volcanic dikes

Augite is a common mineral in the pyroxene group, related to wollastonite, diopside, hedenbergite, pigeonite, and others. Augite typically forms short, stubby crystals in the monoclinic system, but they are usually too small to detect without a hand lens. Large crystals do form occasionally and can be striking and quite shiny, but pit quickly when exposed to air. Augite has a greenish-white streak, which is a good field test.

Augite is commonly found in igneous rocks, sometimes as an essential mineral in dikes. It is common in the Southwest, particularly in Arizona.

# Barite

*Honey-yellow barite crystals from Elko County, Nevada*
PHOTO COURTESY OF THE RICE NORTHWEST MUSEUM OF ROCKS AND MINERALS

**Barium sulfate, BaSO$_4$**
**Family:** Sulfates
**Mohs:** 3–3$\frac{1}{2}$
**Specific gravity:** 4.3–4.6
**Key test(s):** Heavy; fluorescence
**Likely locale(s):** Layered sedimentary rocks; veins with metal ores

Barite usually occurs as pale-yellow to brownish-white tabular crystals. Its color can range from white to dark brown, however, depending on impurities that substitute for the barium ion. Crystal structure in the field can range from slender prisms to fibrous masses. Barite has a white streak and a vitreous, glassy luster. Some barite specimens fluoresce.

Barite is common throughout the Southwest, and most state guidebooks list it. Certain locales feature beautiful crystals. Nevada's Elko County hosts the Meikle Mine, where striking bright-yellow barite clusters have been found. In fact, Nevada is the leading US producer of barite. In Los Angeles County, California, collectors have scoured the Palos Verdes Hills for barite. Arizona's Magma Mine, in Pinal County, is also prominent in the literature, as are veins from the Granite Reef in Maricopa County. The Yellow Cat Mine is covered in *Rockhounding Utah*.

# Benitoite

*Brilliant blue benitoite is the state gemstone of California.*
PHOTO COURTESY OF THE RICE NORTHWEST MUSEUM OF ROCKS AND MINERALS

**BaTi(Si$_3$O$_9$)**
**Family:** Silicates
**Mohs:** 6–6$\frac{1}{2}$
**Specific gravity:** 3.6
**Key test(s):** White streak, fluorescence
**Likely locale(s):** San Benito County, California

Benitoite is California's state gemstone. It is often a striking blue color when of gemstone quality. It has a vitreous luster, a conchoidal fracture, poor cleavage, and hexagonal crystal structure. At the Dallas Gem Mine in the Diablo Range of California, benitoite occurs in natrolite veins found in a glaucophane schist that cuts through a serpentinite zone. That's a unique combination, accounting for this mineral's rarity. Collectors often must dissolve the natrolite host material with acid to release the benitoite crystals for lapidary work. Attractive plates of crystals in their native host material are highly sought as well.

The type locality for this mineral is in San Benito County, California. Tulare, Kern, Fresno, and Mariposa Counties also host deposits. Consult benitoitemining.com for information about their fee-dig operation.

# Beryl

*Common beryl*
PHOTO COURTESY OF THE RICE NORTHWEST MUSEUM OF ROCKS AND MINERALS

$Be_3Al_2Si_6O_{18}$
**Family:** Silicates
**Mohs:** $7\frac{1}{2}$–8
**Specific gravity:** 2.7–2.9
**Key test(s):** Hard; hexagonal form
**Likely locale(s):** Granite pegmatites

Like corundum, beryl comes in a variety. Bright-green beryl is an emerald, while blue beryl is aquamarine. When red or pink, beryl is called morganite, and there is a brilliant red from Utah known as scarlet emerald. Other varieties include a golden beryl, which is yellow, and heliodor, which is greenish yellow. Goshenite is the name for colorless beryl. In all forms beryl has a vitreous luster and a colorless streak. Crystals are hexagonal, sometimes in striking, perfectly six-sided prisms. Cleavage is not a good test, but hardness is, as beryl is harder than quartz. Common beryl is a constituent of many pegmatites in the Southwest, while gem-quality beryl typically occurs in vuggy cavities and pockets in pegmatite.

Red beryl is found at Topaz Mountain in Utah and in Sierra County, New Mexico. Consult *Rockhounding Utah* and *Rockhounding New Mexico* for more information.

# Calcite

*Large calcite crystals in a cluster*

**Calcium carbonate, CaCO₃**
**Family:** Carbonates
**Mohs:** 3
**Specific gravity:** 2.7
**Key test(s):** Rhombohedral angles; hardness
**Likely locale(s):** Veins

Calcite is very common and is found as crystals in pegmatites, as veins in basalts, and mixed with zeolites. Few collectors specialize in calcite, but it's a good mineral to learn. It is usually yellow but can occur in a pure, clear crystal form. It has a white streak, and its luster is vitreous but sometimes very dull. One of the most classic field tests for calcite is the rhombohedral crystal habit, and calcite will cleave easily along those planes. Calcite also fluoresces. Gypsum doesn't effervesce in acid. Two well-known crystal forms are dogtooth and nail head calcite. Perfect calcite crystals display double refraction, in essence doubling the image they are set upon.

Calcite is common to igneous, metamorphic, and sedimentary rocks and is widespread throughout the Southwest. Southern California, Arizona, and Utah all have numerous calcite locales; Socorro, Grant, and Taos Counties in New Mexico are also noteworthy.

# Carnelian

*Bright orange carnelian rough from various locales, ready for the tumbler*

**Quartz, SiO$_2$**
**Family:** Cryptocrystalline quartz
**Mohs:** 6–7
**Specific gravity:** 2.6–2.64
**Key test(s):** Color
**Likely locale(s):** Seams within basalt flows

Carnelian is yet another cryptocrystalline form of quartz, with small amounts of iron giving it a brownish-red tint. Basically, it is an orange form of chalcedony or agate. The term "sard" is used to denote carnelian that is darkly opaque and harder, but both can be present in the same specimen. Carnelian is about as hard as quartz and agate, at 6 to 7 on the Mohs scale, and it is generally softer than sard; both have a white streak. Carnelian also has a characteristic conchoidal fracture pattern. It takes an excellent polish, showing a soft reddish orange.

Carnelian is not common in the Southwest. Some favored locales include the Woodward Ranch near Alpine, Texas; at Cisco, Utah, in Grand County; in the Holbrook District in Navajo County, Arizona; and at Nipomo, in San Luis Obispo County, California.

# Chalcedony

*Chalcedony rose from the southern California desert*

**Quartz, SiO$_2$**
**Family:** Cryptocrystalline quartz
**Mohs:** $6\frac{1}{2}$–7
**Specific gravity:** 2.6–2.64
**Key test(s):** No cleavage; conchoidal fracture
**Likely locale(s):** Quartz-rich areas

MINERALS

Chalcedony is one of the most common forms of quartz, and it is an all-inclusive term for a vast variety of collectible material: agate, bloodstone, heliotrope, chrysoprase, jasper, and flint all qualify. When red or orange, chalcedony is usually referred to as sard, or carnelian. Other varieties include layering, making for sardonyx; green chalcedony with red spots is called bloodstone, or heliotrope. When banded, or clear, rockhounds use the term "agate"; if there are inclusions, the sample is called moss agate. Apple-green chalcedony is called chrysoprase. When red, yellow, or brown, we use the term "jasper," and when white or gray, we call it flint. There is no cleavage, as there is no crystal habit, and the fracture is conchoidal. Chalcedony typically has a waxy, vitreous, or dull luster. Chalcedony is found in virtually all alluvial gravels and beach deposits to some degree. Its various forms are common throughout the Southwest deserts. Refer to *Rockhounding California* for information about chalcedony roses at Augustine Pass, and *Rockhounding Nevada* lists numerous locales across the state.

# Chrysocolla

*Striking blue chrysocolla from the Inspiration Mine in Gila County, Arizona*
PHOTO COURTESY OF THE RICE NORTHWEST MUSEUM OF ROCKS AND MINERALS

**$(Cu,Al)_2H_2Si_2O_5(OH)_4 \cdot nH_2O$**
**Family:** Silicates
**Mohs:** $2\frac{1}{2}$–$3\frac{1}{2}$
**Specific gravity:** 1.9–2.4
**Key test(s):** Soft
**Likely locale(s):** Copper-producing regions

Chrysocolla is a striking blue or green material, usually banded, and softer than turquoise. It produces a streak that is slightly blue green in color, rather than the usual white. Its luster is vitreous to dull, and it forms in botryoidal masses or coatings rather than as crystals or veins. Chrysocolla oxidizes from copper deposits. Its softness precludes it from being an important lapidary material, but some specimens will take a polish.

There are numerous chrysocolla deposits throughout the copper-mining regions of the Southwest. Pinal County, Arizona, produces excellent light-blue specimens; refer to *Rockhounding Arizona* for more information. *Rockhounding New Mexico* and *Rockhounding Nevada* also list numerous chrysocolla locales.

# Corundum

*Low-grade corundum*

**Aluminum oxide, $Al_2O_3$**
**Family:** Oxides
**Mohs:** 9
**Specific gravity:** 3.9–4.1
**Key test(s):** Only thing harder is a diamond
**Likely locale(s):** Alluvial deposits

North American corundum comes in three main varieties—emery, which is black or dark gray and historically used for abrasives; sapphire, which can be blue, yellow, purple, or green; and rubies, which are red. Each variety has a hardness of 9 on the Mohs scale, so all forms of corundum are scratched only by diamond. Luster ranges from vitreous all the way to adamantine, especially in true, gem-quality rubies and sapphires. The hexagonal crystal habit is not a good field indicator, mostly because it is rare to find crystals big enough to inspect. The hardness test is easier in any case.

Riverside County in Southern California's interior mountains contains corundum deposits. Pine Mountain, Utah, and Red Lake, Arizona, are among the few Southwest locales. However, none of the state guidebooks lists locales for collecting corundum.

# Dioptase

*Beautiful blue-green dioptase crystals are sometimes mistaken for emeralds.*
PHOTO COURTESY OF THE RICE NORTHWEST MUSEUM OF ROCKS AND MINERALS

**CuSiO$_3$·H$_2$O**
**Family:** Copper silicates
**Mohs:** 5
**Specific gravity:** 3.3
**Key test(s):** Green streak
**Likely locale(s):** Oxidized copper sulfide zones

Dioptase is not a common mineral. It is a striking blue-green color and could be confused with emeralds if not for the softness—emeralds are a form of corundum and reach 8 on the Mohs scale, while dioptase crystals are at 5. Dioptase has a green streak and vitreous luster and has perfect cleavage in three directions. It is often present as a thin, crystalline coating, so coarser specimens with distinguishable six-sided crystals fetch higher values. Such crystals are often brittle and fragile.

Two deposits in the Southwest are noteworthy: at the Christmas Mine near Hayden, Arizona, and at the Mammoth–St. Anthony Mine near Mammoth, Arizona. There are numerous dioptase occurrences throughout Arizona's copper-producing region; other locales include Clark County, Nevada, and San Bernardino County, California. *Rockhounding Arizona* lists a dioptase-bearing locale on the dumps of the Harquahala Mine.

# Epidote

*Fine epidote crystals from Garnet Hill, California*
PHOTO COURTESY OF THE RICE NORTHWEST MUSEUM OF ROCKS AND MINERALS

$Ca_2Al_2Fe^{3+}OSiO_4Si_2O_7(OH)$
**Family:** Silicates
**Mohs:** 6
**Specific gravity:** 3.3–3.6
**Key test(s):** Color
**Likely locale(s):** Andesite encrustation

Epidote is a common rock-forming mineral, with a characteristic green color, although it can occur as yellow green or even greenish black. Epidote has a colorless to gray streak. Crystals are common in the monoclinic system, forming long, slender crystals or as massively tabular encrustations. Epidote occurs in many situations: in epidote schist; in granite pegmatites; in basalt cavities; in andesite porphyries; in greenstones; and in regional metamorphic rocks.

Epidote is common across the Southwest, ranging from Big Bend in west Texas to Salinas, California, and including Utah, Nevada, New Mexico, and Arizona. *Rockhounding Nevada* contains a locale at Mason Pass; *Rockhounding California* lists an epidote locale in the Marble Mountains, while *Rockhounding New Mexico* lists numerous opportunities.

# Feldspars—Alkali—Microcline

*Feldspars are a key rock-forming mineral, but rarely present as large, identifiable crystals.*
PHOTO COURTESY OF THE RICE NORTHWEST MUSEUM OF ROCKS AND MINERALS

**(KAl, NaAl, or CaAl$_2$)Si$_3$O$_8$**
**Family:** Tectosilicates
**Mohs:** 6–6$\frac{1}{2}$
**Specific gravity:** 2.5–2.6
**Key test(s):** Hardness
**Likely locale(s):** Common rock-forming mineral

Alkali feldspars, sometimes known as potassium feldspars or potash feldspars, are usually white, light gray, or light pink. One exception is amazonite, which appears light blue or green. These varieties of microcline have a vitreous luster and leave a white streak. Feldspars are key rock-forming minerals for igneous and metamorphic rocks and make up more than 50 percent of the Earth's crust. There are semiprecious gem varieties.

Microcline isn't a highly sought mineral, so there are few mentions in the guidebooks. *Rockhounding California* contains two mentions for feldspars. The California-Arizona border region contains numerous microcline locales; Arizona hosts a handful, as do Nevada and New Mexico. The Zapot pegmatite in the Gillis Range of Mineral County, Nevada, is one of the only amazonite locales in the Southwest. Another locale is at Haystack Mountain in Inyo County, California.

# Feldspars–Plagioclase–Orthoclase

*Well-formed orthoclase feldspar crystals*

**Othoclase: (KAl, NaAl, or CaAl$_2$)Si$_3$O$_8$**
**Family:** Tectosilicates
**Mohs:** 6–6½
**Specific gravity:** 2.6–2.8
**Key test(s):** Hardness
**Likely locale(s):** Common rock-forming mineral

Plagioclase feldspars are noted by how much anorthite is present, ranging from zero to 100 percent, from pure albite to pure anorthite. These common rock-forming minerals have a vitreous luster and a white streak. The color is variable but is typically dull white or light gray in appearance, with light tinting common. Feldspars show good cleavage in two directions at ninety degrees, and good twinning. Adularia is transparent. Sunstones are a form of oligoclase feldspar. Labradorite is another gem variety of orthoclase feldspar.

Southern California and western Arizona both contain numerous plagioclase feldspar deposits, usually associated with pegmatites and other minerals.

# Fluorite

*Attractive green fluorite from the Felix Mine, near Azusa in Los Angeles County, California*
PHOTO COURTESY OF THE RICE NORTHWEST MUSEUM OF ROCKS AND MINERALS

**Calcium fluoride, $CaF_2$**
**Family:** Halides
**Mohs:** 4
**Specific gravity:** 3.0–3.2
**Key test(s):** Pastel purple, pink, green color; cubic crystal habit
**Likely locale(s):** Found in hydrothermal, sedimentary, igneous, and volcanic deposits

Fluorite comes in a variety of colors, such as purple, green, and pink, and banding is common. It leaves a white streak, as do many minerals, so a better test is the hardness test or the square crystals. Fluorite crystals are unique in that they cleave perfectly in four directions. Fluorite has a vitreous luster and is softer than quartz but harder than calcite. Another good field test is to check for fluorescence. Fluorite veins are often associated with economic ore deposits, such as galena and sphalerite.

The guidebooks for all Southwest states list fluorite locales. In Texas both Hudspeth County and Brewster County feature locales. New Mexico includes about twenty different fluorite locales, with Socorro County yielding excellent material. California's Los Angeles and San Bernardino Counties have notable showings, such as the Orocopia area. Utah's Deer Trail Mine has yielded excellent specimens in the past; Minersville has seen activity for years. Nevada has had lesser amounts, although the Kaiser Mine has been a big producer. Yuma, La Paz, and Cochise Counties in Arizona have all yielded excellent fluorite specimens.

# Fulgurite

*This fulgurite is from Antelope Valley in Los Angeles County, California.*
PHOTO COURTESY OF THE RICE NORTHWEST MUSEUM OF ROCKS AND MINERALS

**Quartz, SiO$_2$**
**Family:** Quartz
**Mohs:** 7
**Specific gravity:** 2.65
**Key test(s):** Tubelike fused glass; rough exterior
**Likely locale(s):** Sandy deserts

Fulgurites are the result of lightning strikes hitting quartz-rich sand deposits. The intense heat of the strike instantly fuses the quartz in the sand, usually in a tube that is rough on the outside and smooth, sometimes bubbly, on the inside. The resulting color is based on the melted material and is usually gray, light gray, or tan. Fulgurites can extend several feet into the ground, usually branching or tapering away eventually.

None of the guidebooks for this region lists collecting spots, so you're on your own. Look for fulgurites in any flat, open sandy desert or dune area where thunderstorms are common. Mountaintops composed of quartzite are known producers, and exposed areas composed of metal that can attract lightning are good sources as well. Two known areas are in the Plomosa District of La Paz County, Arizona, and in Riverside County, California.

# Geodes

*Typical geodes, also known as thunder eggs. Sliced specimen is from Brewster County, Texas.*
PHOTO (RIGHT) COURTESY OF THE RICE NORTHWEST MUSEUM OF ROCKS AND MINERALS

**Quartz, SiO$_2$**
**Family:** Crystalline quartz
**Mohs:** Varies
**Specific gravity:** Varies
**Key test(s):** Round shape
**Likely locale(s):** Rhyolite lava beds

Geodes are round, quartz-filled structures that range greatly in size and shape. Some are almost perfectly round, about the size of a small fist, and when completely full of quartz, agate, or chalcedony, are usually called thunder eggs. Geodes can be dull and indistinct on the outside and are usually identifiable only by round form, or after breaking open. What appears to create the round shape of a geode is the occurrence of large gas bubbles in molten lava, which may harden and then fill in or replace with quartz-rich ground solutions over long periods of time.

The guidebooks for several Southwest states list geode locales. The Dugway geode beds in Juab County, Utah, are famous worldwide, as are the Hauser geode beds at Wiley Well in northern Imperial Valley, California. There is a productive collecting locale near Payson, Arizona, and in New Mexico collectors frequent Rock Hound State Park near Deming in Luna County.

# Gypsum

*Satin spar variety of gypsum*
PHOTO COURTESY OF THE RICE NORTHWEST MUSEUM OF ROCKS AND MINERALS

**Calcium sulfate, CaSo$_4$·2H$_2$0**
**Family:** Sulfates
**Mohs:** 1½–2
**Specific gravity:** 2.3–2.4
**Key test(s):** Softer than calcite
**Likely locale(s):** Hydrothermal basins

Gypsum is usually white in hand specimens but sometimes appears colorless in pure crystals. It leaves a white streak. Crystals are monoclinic; a rhombic pattern is common, as is massive, fibrous "satin spar" form. Cleavage is perfect in one direction. Gypsum is softer than calcite, which is an easy field test. It is common in hydrothermal replacement deposits and in sedimentary basins, where it is sometimes strip-mined in large quantities for use in sheet rock or as fertilizer. One form of gypsum, selenite, was discovered in a Mexican cave in the form of 30-foot crystals.

There are numerous gypsum deposits across the Southwest, with White Sands National Monument a striking example. *Rockhounding Arizona* lists a selenite rose locale. *Rockhounding Nevada* refers to multiple gypsum sites.

# Halite

*Cubic halite from Southern California*
PHOTO COURTESY OF THE RICE NORTHWEST MUSEUM OF ROCKS AND MINERALS

**Sodium chloride, NaCl**
**Family:** Chlorides
**Mohs:** $2-2\frac{1}{2}$
**Specific gravity:** 2.16
**Key test(s):** Taste
**Likely locale(s):** Evaporite deposits

Halite, or rock salt, is usually clear but is sometimes tinted pink or gray. It leaves a white streak. Halite is strongly isometric and when pure features well-formed crystals in tiny cubes. It has a vitreous luster and perfect cleavage in three directions. Halite is similar to cryolite but harder, and it has a salty taste. Halite is typically found in large evaporite deposits and in large dome-like underground deposits. It mixes readily with other ions, and there are perhaps twenty more halide minerals, such as sulphohalite and polyhalite.

Halite is not a highly sought mineral by collectors, and the guidebooks don't list any sites. Many dried-up bodies of water often contain halite or other salts—Utah's Great Salt Lake is a prime example. California's Searles Lake is another such feature. Zuni Salt Lake in New Mexico, Zimpleman Salt Lake in Hudspeth County, Texas, and the Salt River District in Gila County, Arizona, are three more. A potash mine near Carlsbad Caverns, New Mexico, has yielded striking blue halite crystals.

# Hornblende

*These coarse, black hornblende crystals are from Nevada.*

$(Ca,Na,K)2\text{-}3(Mg,Fe_2+,Fe_3+,Al)_5(SiAl)_8O_{22}(OH)_2$
**Family:** Inosilicates
**Mohs:** 5–6
**Specific gravity:** 3.0–3.4
**Key test(s):** Cleavage angles
**Likely locale(s):** Common rock-forming mineral

Hornblende is a common rock-forming mineral, usually black but sometimes green or brown. It is actually the name given to a series of amphiboles, all differing by their amounts of iron, magnesium, calcium, sodium, aluminum, and potassium. Hornblende leaves a colorless streak, and crystals are monoclinic, usually short, and prismatic. Cleavage is perfect in two directions. It can be confused with schorl, which is black tourmaline, but schorl doesn't cleave like hornblende. Diamond-shaped cleavage angles and a greenish-black color are the best field indicators.

Hornblende is rarely found as collectible crystals, and none of the FalconGuides for the Southwest states contains specific hornblende locales.

# Jasper

*Brown and tan jasper from northern Nevada*

**Quartz, SiO$_2$**
**Family:** Cryptocrystalline quartz
**Mohs:** 7
**Specific gravity:** 2.65
**Key test(s):** Conchoidal fracture
**Likely locale(s):** Basalt flows

Jasper is another common form of quartz, derived from silica-rich ground solutions typically circulating through basalt, which technically makes it a form of chalcedony. Jasper comes in many shades and colors but is usually red, tan, or yellow and sometimes green. It has no crystal habit and leaves a white streak. Gem-quality jasper is hard and shiny, while common jasper is typically porous and won't polish. Jasper is found in most regions where basalt is common, but not always.

Jasper is common in the Southwest. Notable occurrences are found in west Texas in Brewster County, Hudspeth County, and Presidio County. New Mexico's Jornada Draw in Sierra County is important, while Arizona features nearly thirty collectible locales. Utah, Nevada, and Southern California are all known for jasper deposits, so each of the state guidebooks for the Southwest states offers opportunities.

# Jasper—Orbicular

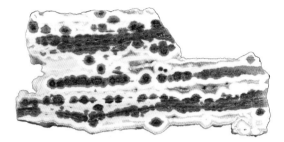

*Note the small round "orbs" that give this jasper its name.*
PHOTO COURTESY OF THE RICE NORTHWEST MUSEUM OF ROCKS AND MINERALS

**Quartz, SiO$_2$**
**Family:** Cryptocrystalline quartz
**Mohs:** 7
**Specific gravity:** 2.65
**Key test(s):** Conchoidal fracture
**Likely locale(s):** Basalt flows

Orbicular jasper is a prized variety of regular jasper that has tiny round inclusions or "orbs" that give it an interesting appearance. A further varietal name is poppy jasper, and yet another is ocean jasper. The orbs are miniature concretions, like in oolitic limestone.

Two notable locales for orbicular jasper are in Arizona's Cochise County at the War Bonnet Ranch, and in California's Santa Cruz Mountains in Paradise Valley. Morgan Hill in Santa Clara County, California, is a noted locale for poppy jasper. *Rockhounding Arizona* lists a locale at Palm Canyon that has yielded good specimens.

# Kyanite

*Kyanite usually occurs as striking blue blades.*
PHOTO COURTESY OF THE RICE NORTHWEST MUSEUM OF ROCKS AND MINERALS

**$Al_2(SiO_4)O$**
**Family:** Silicates
**Mohs:** $5\frac{1}{2}$–7
**Specific gravity:** 3.5–3.7
**Key test(s):** Blue, bladed crystals; splinters easily
**Likely locale(s):** Metamorphic terrain

Increasing metamorphism ⟶

| Chlorite | Biotite | Garnet | Staurolite | Kyanite | Sillimanite |
|----------|---------|--------|------------|---------|-------------|

Kyanite is an indicator mineral that reveals an area has undergone intense regional metamorphism. It is typically blue, sometimes deep blue, and usually found in long, columnar crystals that appear fibrous or bladed. Crystals are often brittle. Kyanite is a great representative of an obscure mineralogical test known as anisotropism. Most minerals don't care how you measure their hardness, but kyanite is an exception. Basically, when you measure the hardness of kyanite, you will measure it at 7 if you go lengthwise along the blade, but only $5\frac{1}{2}$ if you go across the short axis.

Kyanite can be found in metamorphic terrain such as the Pala District in San Diego County, California, and also in that state at the Bluebird Kyanite Mine in Imperial County and the Golden Chariot Mine in Riverside County; it is also associated with pegmatites at Bautista Canyon. The Dome Rock Mountains and Granite Wash Mountains in La Paz County, Arizona, are notable, and there are reports from the Bradshaw Mountains and Phoenix Mountains as well. Kyanite locales aren't listed in the guidebooks for the Southwest states.

# Mica Group–Biotite

*Biotite is a typical mica, usually quite dark.*
PHOTO COURTESY OF THE RICE NORTHWEST MUSEUM OF ROCKS AND MINERALS

**Biotite:** $K(Mg,Fe)_3(Al,Fe)Si_3O_{10}(OH,F)_2$
**Family:** Phyllosilicates
**Mohs:** 2–3
**Specific gravity:** 2.7–3.0
**Key test(s):** Platy cleavage
**Likely locale(s):** Schists

Increasing metamorphism ⟶

| Chlorite | Biotite | Garnet | Staurolite | Kyanite | Sillimanite |
|----------|---------|--------|------------|---------|-------------|

The mica group is composed of several similar sheetlike minerals, including muscovite, biotite, phlogopite, lepidolite, chlorite, and others. Of the two most common varieties, muscovite is white or colorless; biotite is much darker. In fact, mica experts recommend that we no longer use the name "biotite" for a specific chemical formula; instead, they suggest you use "biotite" as a field term for all dark micas you encounter. Micas are found in granitic pegmatites, a historic source of muscovite, while phlogopite occurs in marble and hornfels.

Dark micas that could be called biotite in the field are common throughout the Southwest, usually in schists.

# Mica Group—Chlorite

*This orthoclase feldspar is dusted with a coating of green chlorite.*
PHOTO COURTESY OF THE RICE NORTHWEST MUSEUM OF ROCKS AND MINERALS

$(Mg,Fe)_3(Si,Al)_4O_{10}(OH)_2 \cdot (Mg,Fe)_3(OH)_6$
**Family:** Phyllosilicates
**Mohs:** $2-2\frac{1}{2}$
**Specific gravity:** 2.6–3.3
**Key test(s):** Green, soft; platy cleavage
**Likely locale(s):** Schists; hydrothermal ore deposits

Increasing metamorphism ⟶

| Chlorite | Biotite | Garnet | Staurolite | Kyanite | Sillimanite |
|----------|---------|--------|------------|---------|-------------|

Like biotite, the term "chlorite" no longer refers to a single mineral. Instead, it is the name for a group of related mica minerals, including clinochlore, chamosite, cookeite, and several others. However, field-workers still use the term "chlorite" to refer to green coatings or stains on other minerals. Chlorite represents the lowest grade of metamorphism; it is a key constituent of greenschists, for example. Chlorite has a white streak and is generally not collectible except for the rare forms.

Chlorite is common throughout the low-grade metamorphic rocks of the Southwest.

# Mica Group—Muscovite

*Muscovite sometimes occurs in large sheets called "books."*
PHOTO COURTESY OF THE RICE NORTHWEST MUSEUM OF ROCKS AND MINERALS

**$KAl_2(AlSi_3O_{10})(F,OH)_2$**
**Family:** Phyllosilicates
**Mohs:** 2–2½ parallel to main cleavage, 4 at right angle
**Specific gravity:** 2.8–3.0
**Key test(s):** Platy cleavage; pearly luster
**Likely locale(s):** Pegmatites and schists

Increasing metamorphism ⟶

| Chlorite | Muscovite | Garnet | Staurolite | Kyanite | Sillimanite |
|----------|-----------|--------|------------|---------|-------------|

Muscovite is the most common mica seen in the field and is usually white or colorless; its cousin biotite is much darker. Chlorite is greenish, while lepidolite is pink or light purple. Muscovite has a vitreous, pearly luster, leaves a colorless streak, and perhaps most characteristic, its cleavage is perfect in one direction, resulting sometimes in great sheets called "books." Because muscovite flakes can look yellow and shiny, they are sometimes mistaken for fool's gold, but a knifepoint will fracture mica flakes, while flattening or cutting gold. If large enough, mica will show a hexagonal pattern. Muscovite bends the easiest of the mica group.

Muscovite is common across the Southwest. *Rockhounding Utah* lists a locale at Painter Spring, but the rest of the guidebooks are light on specific muscovite collecting opportunities.

# Olivine

*Olivine-rich basalt*

$(Mg,Fe_2+)2SiO_4$
**Family:** Nesosilicates
**Mohs:** $6\frac{1}{2}-7$
**Specific gravity:** 3.5–4.0
**Key test(s):** White streak
**Likely locale(s):** Found with metamorphic rocks and in hydrothermal replacement deposits

Olivine is usually dark green, with a vitreous luster and colorless streak. When found in crystalline form, such as the semiprecious gemstone peridot, olivine will cleave in two directions, but it is more commonly associated with olivine-rich basalt. Olivine is the name for a series of minerals, with forsterite on the magnesium-rich end of the series and by far the most common, and fayalite on the iron-rich end. Olivine is actually a common rock-forming mineral, composing much of the Earth's mantle.

None of the FalconGuides for the southwestern states lists specific olivine sites, but there are some collecting opportunities. The Wind Mountains and Caballo Mountains of New Mexico contain olivine. Arizona features numerous locales, such as near Prescott, in the Ash Peak District in Greenlee County, and in the San Carlos Peridot Ridge area. In Nevada the Bunkerville District in Clark County has reported olivine. California offers olivine in the Pala District, in the Santa Lucia Mountains, and in the Mesa Grande District, among other areas.

# Onyx

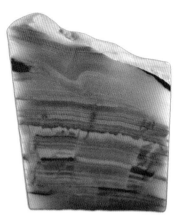

*Polished onyx*
PHOTO COURTESY OF THE RICE NORTHWEST MUSEUM OF ROCKS AND MINERALS

**Quartz, SiO$_2$**
**Family:** Quartz
**Mohs:** 6–7
**Specific gravity:** 2.6–2.7
**Key test(s):** Patterns
**Likely locale(s):** Silica-rich environments

Onyx is yet another form of the chalcedony family of quartz. It is usually uniformly banded, creating striking patterns of red, white, black, or brown. If the bands contain sard, the result is termed "sardonyx." The most common color is black, but rockhounds often label as onyx material formed in white carbonates such as limestone, marble, and calcite. Mineralogists restrict onyx to a variety of agate. The luster is vitreous to silky, streak is white, and onyx is usually slightly translucent.

Mindat.org lists a single true onyx deposit in the southwestern states at the Stout onyx claim in Utah County, Utah. However, there are many more opportunities—*Rockhounding Arizona* lists a locale at Kingman, and *Rockhounding Nevada* refers to a plentiful locale near Jackpot.

# Opal

*Common opal from within basalt flows*

**Quartz, $SiO_2 \cdot nH_2O$**
**Family:** Cryptocrystalline quartz
**Mohs:** $5\frac{1}{2}–6\frac{1}{2}$
**Specific gravity:** 2.0–2.2
**Key test(s):** Softness, lighter than other quartz
**Likely locale(s):** Quartz-rich areas

Opal is another form of quartz, but this time with considerable water present. There are numerous varieties of common opal, with mindat.org showing at least 155 names. Common opal, sometimes referred to as hyalite, has a distinctive vitreous, pearly luster. It is usually white, but colors typically include yellow, brown, green, blue, or gray. Its streak is white. Opal is not crystalline; there is no habit or cleavage, and it has a conchoidal fracture. Opal is softer than agate, jasper, and other forms of quartz. Opal frequently forms in cavities, fractures, and air bubbles in basalt, where it sometimes weathers out and remains as small, round pea-size blebs. Wood and shells often become "opalized" through replacement. Opal will fluoresce, which also sets it apart from jasper and quartz.

Nevada, Utah, Southern California, and Arizona all host numerous opal showings, referenced in the guidebooks for those states. West Texas features the Quitman Mountains and Sierra Blanca Peaks in Hudspeth County. New Mexico features opal deposits at Jornada Draw and in the San Mateo Mountains, among others.

# Orpiment and Realgar

*Red orpiment and yellow realgar from Twin Creeks Mine, Winnemucca, Nevada*
PHOTOS COURTESY OF THE RICE NORTHWEST MUSEUM OF ROCKS AND MINERALS

$As_2O_3$ and $As_4S_4$
**Family:** Arsenides
**Mohs:** $1\frac{1}{2}$–2
**Specific gravity:** 3.5–3.6
**Key test(s):** Orpiment is bright yellow; realgar is deep red
**Likely locale(s):** Hot springs and fumaroles

These two arsenic minerals usually show up with each other; orpiment is an oxide of arsenic, and realgar is a sulfide. Orpiment has a bright-yellow streak and was used as a pigment in ancient times, while realgar is deep orange red. Orpiment is often a by-product of weathered, altered realgar, where the sulfur leaches away, leaving an oxide. Likely collecting locales include hot springs and hydrothermal vein systems. Orpiment is rarely seen as a crystal, and is much more common as a fibrous, velvety coating, sometimes in the botryoidal habit. Realgar is often observed as small, granular crystals, so larger specimens are highly sought.

Nevada, Southern California, and Utah seem to offer the best collecting opportunities for arsenic minerals, but none show up in the guidebooks, as most collecting occurred deep underground. San Benito County, California, has yielded realgar specimens in the past, and Mercur, Utah, and Manhattan, Nevada, are noted in the literature. The Twin Creeks Mine in Humboldt County, Nevada, is known for nice orpiment specimens.

# Petrified Wood

*Oligocene age petrified palm wood from the Catahoula Formation of Fayette County, Texas*

**Quartz, SiO$_2$**
**Family:** Silicates
**Mohs:** 7
**Specific gravity:** 2.2–2.6
**Key test(s):** Rings, lines, and wood features
**Likely locale(s):** Volcanic tuff deposits, sedimentary rocks

Petrified wood is the common name for fossil wood and is used interchangeably with opalized wood, agatized wood, and silicified wood. Petrified wood comes in a wide variety of colors and conditions. It is usually black, red, white, or yellow. Some samples are very hard and take an excellent polish, but other samples may be porous and unsuitable for jewelry.

Petrified wood is common throughout the Southwest, and every guidebook lists at least one collecting locale. At Virgin Valley, Nevada, rockhounds have unearthed opalized wood and bone with fantastic rainbow colors that can unfortunately dry out quickly. Other famous locales for the more common petrified wood include Petrified Forest National Park in Arizona and Escalante Petrified Forest in Utah; collecting isn't allowed in the parks, but there are locales nearby. There is a noteworthy palm wood collecting site in Fayette County, Texas.

# Quartz—Crystalline

*Water-clear quartz specimen with excellent crystal faces*

**Quartz, SiO$_2$**
**Family:** Silicates
**Mohs:** 7
**Specific gravity:** 2.65
**Key test(s):** Color
**Likely locale(s):** Seams within basalt flows; pegmatites

Quartz crystals are distinctly hexagonal, with pyramidal terminations. They are most prized when displaying water-clear color and perfect pointed ends. Like most quartz, the streak is white. There is no cleavage in quartz crystals, which have a vitreous, greasy luster. Crystalline quartz is softer than corundum, topaz, and diamond but harder than calcite. Gem-quality quartz crystals are associated with granite pegmatites, but common quartz is abundant and found in all three rock classes.

There is a unique locale of double-terminated, clear quartz crystals known as Pecos Diamonds, covered in *Rockhounding New Mexico*. West Texas offers several quartz-collecting locales, primarily in Brewster and Hudspeth Counties. Utah, Nevada, and California host numerous opportunities; *Rockhounding Arizona* is a good source.

# Quartz–Smoky

*Smoky quartz can range from light brown to jet-black in color.*
PHOTO COURTESY OF THE RICE NORTHWEST MUSEUM OF ROCKS AND MINERALS

**Crystalline quartz, $SiO_2$**
**Family:** Silicates
**Mohs:** 7
**Specific gravity:** 2.65
**Key test(s):** Hardness; glossy black color
**Likely locale(s):** Pegmatites

Smoky quartz is a semiprecious gem variety of common crystalline quartz, so it shares the same diagnostic properties. The hardness is 7, there is a white streak if you can produce it, and crystals are hexagonal, with no cleavage. Collectors prize the specimens with a characteristic pointed, pyramidal termination. The black coloring appears to be the result of free silicon atoms that have been irradiated by the natural background radiation common to granite. Gem-quality smoky quartz crystals are associated with granite pegmatites, often accompanied by feldspar.

Utah's Mineral Mountains and Bovine Mountains report smoky quartz. Arizona's Mineral Mountain District offers numerous opportunities. The Zapot pegmatite in Nevada is a known producer. California's San Diego County area has more than thirty locales that contain at least incidental smoky quartz. Three districts in New Mexico report smoky quartz: Hansonburg, Picuris, and Capitan.

# Rhodochrosite

*The Alma Rose, from the Sweet Home Mine in Colorado*
PHOTO COURTESY OF THE RICE NORTHWEST MUSEUM OF ROCKS AND MINERALS

MINERALS

**Manganese carbonate, $MnCO_3$**
**Family:** Carbonates
**Mohs:** $3\frac{1}{2}$–4
**Specific gravity:** 3.7
**Key test(s):** Deep pink to rose red; white streak, vitreous luster
**Likely locale(s):** Vein mineral in silver districts

Rhodochrosite is a lovely rosy red in its pure form, but impurities typically temper that vivid color down to pink, cinnamon, or even yellow. Its rhombohedral crystals are rare and highly collectible, but most collectors encounter field deposits as ribbons or banded minerals. These specimens will often take a polish and produce striking red-pink, banded lapidary creations. As crystals, rhodochrosite is soft and difficult to fashion into jewelry, and it has perfect cleavage, another strike against it for faceting.

Alma, Colorado, boasts some of the best rhodochrosite in the world. There are numerous minor locales scattered across the Southwest, including Kern County, California; Santa Cruz County, Arizona; and Nye County, Nevada. *Rockhounding New Mexico* lists sites at Questa and at the Luis Lopez manganese district.

# Rhodonite

*Light pink color is characteristic of rhodonite, as are the dark manganese "roads" throughout.*
PHOTO COURTESY OF THE RICE NORTHWEST MUSEUM OF ROCKS AND MINERALS

**Manganese silicate, MnSiO$_3$**
**Family:** Metal silicates
**Mohs:** $5\frac{1}{2}$–$6\frac{1}{2}$
**Specific gravity:** 3.6–3.8
**Key test(s):** Pink (when fresh), turns brown
**Likely locale(s):** Found with metamorphic rocks and in hydrothermal replacement deposits

Rhodonite is usually easy to identify because it is rose pink in the field and is one of the few minerals found in that color. Some rhodonite samples are red, brown, or yellow, depending on the calcium, iron, or magnesium impurities present with the manganese, so the color test isn't foolproof. Most rhodonite samples are massive and pink, with "road-like" black tracking that serves as a memory aid—roads in rhodonite.

There are rhodonite deposits scattered across the Southwest, but none are featured in the FalconGuides. In Utah the Drum Mountains in Juab County host the Manganese King Mine, and rhodonite is reported from the Bingham Canyon Mine in Salt Lake County. The Groundhog Mine in New Mexico's Grant County has rhodonite deposits, as does the Questa Molybdenum Mine in Taos County. Arizona lists numerous rhodonite deposits, as does Southern California.

# Sard

*Sard is almost opaque and deep red, while carnelian is more translucent and orange.*

**Quartz, SiO$_2$**
**Family:** Quartz
**Mohs:** 6–7
**Specific gravity:** 2.60–2.65
**Key test(s):** Harder than carnelian; less orange; more jagged fracture
**Likely locale(s):** Agate locales

Like carnelian, sard is a cryptocrystalline form of quartz, but it appears to contain larger amounts of iron, giving it a dark-brownish/blood-red tint. Sard is thus a special form of carnelian and yields striking cabochons and other jewelry. While sard is typically darker and harder than carnelian, both specimens can be present in the same vein, typically as alternating bands. Carnelian is as hard as quartz and agate, at 7 on the Mohs scale, and it has a white streak. Carnelian also has a characteristic conchoidal fracture pattern, whereas sard breaks with a more jagged, hackled pattern.

One noteworthy locale is at Black Hills in the Palo Verde District in Imperial County, California. The Old Laguna claims in Cibola County, New Mexico, also rate a mention. *Rockhounding New Mexico* is an excellent resource.

# Selenite

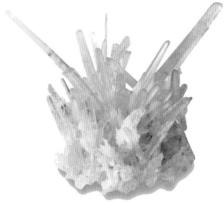

*Gypsum crystals from "Cave of Swords" in Chihuahua, Mexico*
PHOTO COURTESY OF THE RICE NORTHWEST MUSEUM OF ROCKS AND MINERALS

$CaSO_4 \cdot 2H_2O$
**Family:** Sulfates
**Mohs:** 2
**Specific gravity:** 2.3
**Key test(s):** Water-clear crystals
**Likely locale(s):** With gypsum

Selenite is a form of gypsum and is a general term for satin spar, desert rose, gypsum flower, and selenite crystals. It does not contain the element selenium. Rockhounds typically reserve the name selenite for superior, transparent gypsum crystals found either individually or as masses. Selenite comes in a variety of colors, including brownish green, brownish yellow, gray, white, and other light tints. Primarily, however, think of the most water-clear tabular and columnar crystals as being selenite. Selenite has a white streak and a pearly luster.

Selenite is common across the southwestern states. Utah's Blue Hills also contain collecting locales, listed in *Rockhounding Utah*. The Kernick Mine, listed in *Rockhounding Nevada*, is another good source.

# Septarian Nodules

*Septarian nodule from Muddy Creek, near Orderville in Kane County, Utah*
PHOTO COURTESY OF THE RICE NORTHWEST MUSEUM OF ROCKS AND MINERALS

**Formula:** $CaCO_3$, with $FeS_2$, $SiO_2$, or $BaSO_4$ (calcium carbonate with pyrite, quartz, or barite)
**Family:** Varied
**Mohs:** Varied
**Specific gravity:** Variable, usually around 2.6
**Key test(s):** Takes a polish
**Likely locale(s):** Sedimentary rocks

Septarian nodules are like "fossilized concretions." They are highly prized by collectors because they take a nice polish, feature dramatic mineral-filled cracks, and have an unusual, rounded shape. They also represent an interesting geologic background that "rock detectives" may or may not have puzzled out, involving cracking and filling, circulating groundwater, heat and pressure, and perhaps multiple chemical processes.

Alton, Utah, hosts a good collecting locale for septarian nodules, as does nearby Mount Carmel; both are listed in *Rockhounding Utah*.

# Sillimanite

*Various rounded sillimanite pebbles; this mineral is difficult to locate as pure crystals.*

**$Al_2SiO_5$**
**Family:** Nesosilicates
**Mohs:** 7
**Specific gravity:** 3.2
**Key test(s):** Silky, fibrous
**Likely locale(s):** Severely metamorphosed aluminum-rich sedimentary rocks

Increasing metamorphism ⟶

| Chlorite | Biotite | Garnet | Staurolite | Kyanite | Sillimanite |
|----------|---------|--------|------------|---------|-------------|

Sillimanite is an indicator mineral for rocks that have undergone metamorphism at the highest temperature. In other words, rocks with sillimanite got metamorphosed more than rocks with kyanite—even though the two minerals share the same formula (along with andalusite). Sillimanite crystals are usually rare and small; it is more common to find rocks with lots of sillimanite present, and they sometimes appear to shimmer. This is due to the silky luster, the fibrous mass of crystals, and the wavy nature of their alignment.

Sillimanite occurs in Southern California and in Arizona but does not appear in any of the guidebooks.

# Staurolite

*Staurolite schist, left, and staurolite twin from near Taos, New Mexico*
PHOTO (RIGHT) COURTESY OF THE RICE NORTHWEST MUSEUM OF ROCKS AND MINERALS

$Fe_2Al_9Si_4O_{22}(OH)_2$
**Family:** Nesosilicates
**Mohs:** $7–7\frac{1}{2}$
**Specific gravity:** 3.7–3.8
**Key test(s):** Hardness, twinning
**Likely locale(s):** Highly metamorphosed schists

Increasing metamorphism ⟶

| Chlorite | Biotite | Garnet | Staurolite | Kyanite | Sillimanite |
| --- | --- | --- | --- | --- | --- |

Staurolite, which is mainly found in staurolite schist, is an indicator mineral for intense metamorphism. A staurolite schist denotes a regionally metamorphosed rock between garnet and kyanite grade. Staurolite crystals are usually brown, with a vitreous luster and a white streak. The crystal habit is monoclinic; twinning is common, such as right-angle "fairy" crosses. Cleavage is poor, in lengthwise direction.

The most popular collecting locale for staurolite in the Southwest is south of Taos, New Mexico, and *Rockhounding New Mexico* lists a site in the Picuris Mountains. Arizona features several locales, including the Bradshaw Range, the Wickenburg Mountains, and the Midnight Owl pegmatite, all in Yavapai County, and the Piestewa Peak area. Vitrefax Hill in Imperial County, California, is also mentioned in the literature, as is Beatty, Nevada.

# Talc

*Talc is so soft, it is easy to carve with your fingernail.*
PHOTO COURTESY OF THE RICE NORTHWEST MUSEUM OF ROCKS AND MINERALS

$Mg_3Si_4O_{10}(OH)_2$
**Family:** Magnesium silicates
**Mohs:** 1
**Specific gravity:** 2.7–2.8
**Key test(s):** Soft
**Likely locale(s):** Metamorphic terrain

Talc is easy to test in the field because you can scratch it with your fingernail—it represents 1 on the Mohs hardness scale. It is the main component of most soapstones and is usually found as a mix of magnesium silicates, such as tremolite or magnesite. Beware of the tendency for asbestos to accompany talc. Color varies, depending on associated minerals present, and can range from white to bright green. The streak is white. Talc has a greasy feel and shows a pearly luster. Crystals are extremely rare. Look for talc in metamorphic regions with ultramafic rocks such as dunite or peridotite.

Talc doesn't show up in the FalconGuides, but it is found throughout the Southwest. One prime area for talc is the belt of metamorphic rocks in western California extending north from Santa Barbara. Hudspeth County, Texas, contains the Alla-moore Talc District. Southeastern Arizona contains numerous talc deposits.

# Tourmaline–Schorl

*Schorl often forms compact masses of parallel crystals.*
PHOTO COURTESY OF THE RICE NORTHWEST MUSEUM OF ROCKS AND MINERALS

$(Ca,K,Na)(Al,Fe,Li,Mg,Mn)_3(Al,Cr,Fe,V)_6(BO_3)_3(Si,Al,B)_6O_{18}(OH,F)_4$

**Family:** Cyclosilicates
**Mohs:** $7–7\frac{1}{2}$
**Specific gravity:** 3.0–3.3
**Key test(s):** Harder than apatite
**Likely locale(s):** Pegmatites

Tourmaline comes in a variety of forms, with the most common being black schorl. That's by far the most common tourmaline species, but semiprecious gem varieties include elbaite, indicolite, rubellite, and dravite. Tourmaline has a vitreous luster, leaves a white streak, and forms a hexagonal crystal with no cleavage. Hornblende is also black, but schorl has a triangular cross-section.

Southern California offers more than 200 mentions of tourmaline in the literature; Arizona shows around seventy. Hudspeth County, Texas, offers several tourmaline locales. The Harding pegmatite in the Picuris District, Taos County, New Mexico, is notable. Consider also Utah's Mercur District in Tooele County and the Newberry Mountains in Clark County, Nevada. *Rockhounding California* and *Rockhounding Mew Mexico* are your best starting points for tourmaline collecting locales.

# Variscite

*Fine variscite specimen from Utah County, Utah*
PHOTO COURTESY OF THE RICE NORTHWEST MUSEUM OF ROCKS AND MINERALS

**$AlPO_4 \cdot 2H_2O$**
**Family:** Phosphates
**Mohs:** $4\frac{1}{2}$
**Specific gravity:** 2.6
**Key test(s):** Greenish; veined
**Likely locale(s):** Phosphate regions

Variscite is a rare secondary mineral formed when phosphate-rich groundwater leaches through aluminum-rich rocks. It usually forms as attractive greenish-blue masses or nodules, frequently shot through with white, brown, or black veins. It is lighter than malachite and greener than turquoise, two minerals with which it can be confused. Variscite has a white streak, a vitreous to waxy luster, and takes a nice polish. There are many known varieties of this mineral, depending on the ratio of aluminum and iron, the presence of arsenic, and the amount of other elements available.

The guidebooks for Southwest states don't show variscite locales open to the public. The two most prevalent producers of variscite are Lander County, Nevada, and Utah County, Utah. The Bisbee District of Arizona has produced variscite as well.

# Zeolite Group: Heulandite

*Nice pink heulandite specimen*
PHOTO COURTESY OF THE RICE NORTHWEST MUSEUM OF ROCKS AND MINERALS

$(Ca,Na)(Al_2Si_7O_{18})\cdot 6H_2O$
**Family:** Zeolites
**Mohs:** $3–3\frac{1}{2}$
**Specific gravity:** 2.2
**Key test(s):** Pink (when fresh)
**Likely locale(s):** Basalt flows and hornfels

"Heulandite" is the name for a group of zeolites known for a distinctive, pink-colored form, but often crystals are white or lack color completely. The streak is white. Crystals are monoclinic and sometimes display a unique coffin shape. Luster can be pearly or vitreous. Heulandite is usually found with other zeolites in veins or pockets within basalt and andesite but can also occur in schists.

Heulandite occurs in Southern California at the Himalaya Mine, in the Rincon District, and in the Pala District in San Diego County. In Arizona, Malpais Hill, Alum Mountain, the Little Ajo Mountains, and the New Water Mountains are all sources for heulandite.

# Zeolite Group: Mordenite

*Short, soft, needlelike crystals are typical for mordenite.*
PHOTO COURTESY OF THE RICE NORTHWEST MUSEUM OF ROCKS AND MINERALS

$(Ca, Na_2, K_2)Al_2Si_{10}O_{24} \cdot 7H_2O$
**Family:** Zeolites
**Mohs:** 3–4
**Specific gravity:** 2.1
**Key test(s):** Fibrous
**Likely locale(s):** Veins and cavities in basalt flows

Mordenite forms sprays of delicate needles, which makes it one of the easier zeolites to identify. It is usually white, although it can discolor quickly if it begins to absorb moisture. It streaks white or colorless and usually forms as needlelike crystals but can also form velvety coatings. It is usually found in the cavities and vugs of volcanic rocks.

The Crestmore quarries in Riverside County, California, have yielded mordenite in the past. Tick Canyon and Newbury Park in Ventura County are also noted in the literature. In Arizona the Oatman District, New Water District, Cottonwood Basin, and Copper Mountain District all receive mention. The Sugarland Tuff in Luna County, New Mexico, also contains mordenite.

# Zeolite Group: Scolecite

*Scolecite crystals form long, delicate sprays.*
PHOTO COURTESY OF THE RICE NORTHWEST MUSEUM OF ROCKS AND MINERALS

$CaAl_2Si_3O_{10} \cdot 3H_2O$
**Family:** Zeolites
**Mohs:** $5-5\frac{1}{2}$
**Specific gravity:** 2.3
**Key test(s):** Crystal spray
**Likely locale(s):** Basalt flows and hornfels

Like mordenite, scolecite displays thin, needlelike crystals, but in scolecite, the needles are thicker and stronger, rather than delicate. Scolecite is a common zeolite, usually colorless or white, but some forms are light pink, and delicate shades of red and green show up occasionally. It is difficult to distinguish from the zeolite natrolite—a detailed optical examination is needed.

There are two locales listed on mindat.org for scolecite in the Southwest: in California at the Crestmore quarries in Riverside County, and in the Lone Star District, which is on the San Carlos Indian Reservation and off-limits to collectors.

# Zeolite Group: Stilbite

*Stilbite crystals are usually short, stubby, and milky white.*
PHOTO COURTESY OF THE RICE NORTHWEST MUSEUM OF ROCKS AND MINERALS

$NaCa_2Al_3Si_{13}O_{36} \cdot 16H_2O$
**Family:** Zeolites
**Mohs:** $3\frac{1}{2}$–4
**Specific gravity:** 2.12–2.22
**Key test(s):** Pearly, vitreous luster
**Likely locale(s):** Basalt flows and hornfels

Like heulandite, stilbite is the name for a series of zeolites, rather than a single species, but the difference is in the ratio of sodium and calcium present. Most stilbites are richer in calcium. Crystals are usually colorless or white, but they can be pink on occasion. Iceland has long been a prime resource for excellent stilbite specimens, but stilbite is a common zeolite and is found alongside the rest of the zeolite family in vugs and cavities within zeolite-rich basalts.

Stilbite is fairly common in California and Arizona and shows up where mordenite and scolecite are found. In addition, stilbite is reported from Walnut Draw in the Davis Mountains of Brewster County, Texas; in the Preuss District of Beaver County, Utah; and at Garnet Hill near Ely and the Getchell Mine near Adam Peak in Nevada, among many other places.

# Metallic Minerals

## Arsenopyrite

*Arsenopyrite crystals have a warped look compared to pyrite.*
PHOTO COURTESY OF THE RICE NORTHWEST MUSEUM OF ROCKS AND MINERALS

**FeAsS**
**Family:** Metal sulfides
**Mohs:** $5\frac{1}{2}$–6
**Specific gravity:** 5.9–6.2
**Key test(s):** Density; garlic odor when crushed
**Likely locale(s):** Pegmatites, vein systems

Arsenopyrite is a variation on common pyrite, with an arsenic ion substituting for iron in the crystal lattice. That substitution tends to warp the crystals—where pyrite has more striking angles and lines, arsenopyrite is frequently striated and twinned. Where pyrite is a yellow brass color, arsenopyrite is often more of a silver-white or gray. There are at least five other variations when other ions substitute for the arsenic ion, with cobalt, nickel, stibnite, iridium, and others present in varying ratios. Arsenopyrite is not very stable and tends to oxidize. Interestingly, this variant on fool's gold can be a key indicator to the presence of large amounts of gold.

Arsenopyrite is common in the Southwest, but there are no specific locales listed in the guidebooks.

# Azurite

*Velvety azurite from Greenlee County, Arizona*
PHOTO COURTESY OF THE RICE NORTHWEST MUSEUM OF ROCKS AND MINERALS

**Copper carbonate, $Cu_3(CO_3)_2(OH)_2$**
**Family:** Metal carbonates
**Mohs:** $5\frac{1}{2}$–6
**Specific gravity:** 3.5–4.0
**Key test(s):** Color; hardness
**Likely locale(s):** Rich copper deposits

Azurite is a striking blue mineral closely associated with green malachite in copper-mining regions. The two make a pleasing combination when fresh, but azurite tends to lose its brilliance if subjected to heat, bright light, or simply too much air over too long a time. When fresh it has a bright-blue streak, which was exploited for paint pigments during the Middle Ages in Europe. Color and streak are the two key characteristics when identifying field samples.

Perhaps the most well-known locale for azurite in the Southwest is at Bisbee, Arizona. There are hundreds of azurite locales in Arizona alone. The Magdalena District in Socorro County, New Mexico, has also offered up excellent specimens. Most known copper regions will host a lot of malachite and a little azurite; Utah's Rock Range Mines are a good place to collect. *Rockhounding Arizona* and *Rockhounding Nevada* are good starting points for collecting.

# Bornite

*Bornite is sometimes called the "peacock ore" due to its many colors.*

**Copper iron sulfide, $Cu_5FeS_4$**
**Family:** Metal sulfides
**Mohs:** $3–3\frac{1}{4}$
**Specific gravity:** 4.9–5.3
**Key test(s):** Peacock color, heft
**Likely locale(s):** Rich copper deposits

Bornite is known as the "peacock" ore because it has so many red, purple, bronze, blue, and other hues, all calling for attention at once. It is an important sulfide of copper because it is rich in copper by weight. Bornite has a metallic luster, and its streak is grayish black, but you probably won't need to streak it for identification purposes thanks to the striking play of colors. Bornite crystals are rare, as they generally appear as disseminated deposits in skarns, veins, and massive sulfide deposits.

Bornite is mentioned specifically at several locales in *Rock-hounding New Mexico*. There are some 250 occurrences of bornite in Arizona, as would befit a state with such an important history of copper production. Culberson County, Texas, has a handful of bornite locales, as do the Burro Mountains of New Mexico. Nevada, Utah, and Southern California all have reported bornite deposits.

# Chalcocite

*This chalcocite specimen is coated with bornite.*
PHOTO COURTESY OF THE RICE NORTHWEST MUSEUM OF ROCKS AND MINERALS

**Copper sulfide, Cu$_2$S**
**Family:** Metal sulfides
**Mohs:** 2½–3
**Specific gravity:** 5.5–5.8
**Key test(s):** Dense; black to lead-gray streak
**Likely locale(s):** Known copper areas

Chalcocite is another important copper ore. It is rarely found as crystals, but when it is, it forms typically in hydrothermal vein systems, with a metallic luster and in tabular form. Chalcocite is much more common as a secondary mineral in massive oxidized zones, where the copper has leached out from other minerals. It is related to at least nine other copper-sulfide minerals that have varying ratios of copper and sulfur; these are known as the Chalcocite-Digenite group.

There are more than a dozen districts across Southern California with significant chalcocite deposits. Nevada, Utah, New Mexico, and west Texas also contain many chalcocite locales, while Arizona has hundreds of mines or prospects containing chalcocite.

# Chalcopyrite

*Cluster of chalcopyrite crystals*
PHOTO COURTESY OF THE RICE NORTHWEST MUSEUM OF ROCKS AND MINERALS

**Copper iron sulfide, $CuFeS_2$**
**Family:** Metal sulfides
**Mohs:** $3\frac{1}{2}$–4
**Specific gravity:** 4.1–4.3
**Key test(s):** Crystals aren't as cubic
**Likely locale(s):** Known copper areas

Chalcopyrite is a common sulfide that is chemically quite similar to common pyrite, except chalcopyrite contains both copper and iron, whereas pyrite has no copper. Like pyrite, chalcopyrite is brassy and golden yellow, but chalcopyrite is softer than pyrite, and its crystal habit is a tetrahedron, while pyrite is cubic. Its streak is greenish black, also similar to pyrite. There is an entire family of minerals related to chalcopyrite, with varying amounts of silver and gallium substituting for iron or copper, and selenium substituting for sulfur. Chalcopyrite is often associated with hydrothermal economic ore deposits that host silver and gold. It is usually found in massive sulfide zones with other sulfides, especially pyrite, and is a primary copper ore.

Most of the metal sulfide mines and prospects in the Southwest that contain pyrite also contain chalcopyrite, especially with copper being so prominent in this area. Arizona has at least 200 locales where chalcopyrite is significant, but *Rockhounding New Mexico* is the best place to start to sleuth out collecting opportunities.

# Cinnabar

*Blood-red cinnabar (left) and silicified cinnabar, known as myrickite (right)*
PHOTO (RIGHT) COURTESY OF THE RICE NORTHWEST MUSEUM OF ROCKS AND MINERALS

**Mercury sulfide, HgS**
**Family:** Metal sulfides
**Mohs:** $5\frac{1}{2}$–6
**Specific gravity:** 3.5–4.0
**Key test(s):** Crimson, pink (when fresh), hard
**Likely locale(s):** Calderas, limestones

Cinnabar is fairly easy to identify in the field, being a vivid red. Hexagonal crystals are quite rare; instead, you should look for reddish streaks and crusts in sulfide-rich zones. If there is enough of a sample for a streak test, cinnabar leaves a scarlet streak. In rich mercury mines with heavy concentrations of cinnabar, miners have actually noted liquid mercury deposits in small pools and puddles, so look for that. Look for cinnabar near hydrothermal zones, such as hot springs, or in big opalized calderas. When agatized or opalized, cinnabar deposits are named myrickite, found in Death Valley and at the Manhattan Mine in Napa County, California.

Southern California, New Mexico, and Utah all list numerous cinnabar deposits in their inventory, and the collecting guides for those states are good places to start. The Cedar City Iron Mines in Utah still supply cinnabar samples in the tailings. Arizona counts at least forty cinnabar deposits in its inventory. The Big Bend Country of west Texas contains at least two dozen cinnabar locales scattered across Brewster County. New Mexico lists the Blanchard Mine in Socorro County.

# Copper

*Native copper from Arizona, still shiny and unspoiled by oxidation*
PHOTO COURTESY OF THE RICE NORTHWEST MUSEUM OF ROCKS AND MINERALS

**Copper, Cu**
**Family:** Elemental metal
**Mohs:** 2½–3
**Specific gravity:** 8.9
**Key test(s):** Shiny as a penny; heavy; poundable
**Likely locale(s):** Found with metamorphic rocks and in hydrothermal replacement deposits

Native copper, when fresh, can appear as bright as a shiny new penny. However, copper quickly tarnishes when exposed to air and starts to turn black, green, or blue. Native copper nuggets leave a characteristic copper streak and are pretty easy to identify without much practice thanks to the copper coins in our financial system. Crystals are rare and are in the isometric habit. Copper is noted for its malleability, meaning it bends and can be pounded, flattened, or rolled.

The Warren District in the Bisbee area of Arizona is one of the most famed locales in the Southwest states for supplying superb native copper specimens. Native copper occurs in several other places in Arizona, including the Ajo, Mineral Creek, Clifton-Morenci, Globe, Bagdad, Safford, and Jerome districts. Fine, branched crystal structures occur in the Mountain City District of Elko County, Nevada.

# Cuprite

*Cuprite, variety chalcotrichite, from the Detroit Shaft in the Morenci District, Greenlee County, Arizona*
PHOTO COURTESY OF THE RICE NORTHWEST MUSEUM OF ROCKS AND MINERALS

**$Cu_2O$**
**Family:** Oxides
**Mohs:** $3\frac{1}{2}$–4
**Specific gravity:** 6.1
**Key test(s):** Dark-red crystals; streak
**Likely locale(s):** Copper regions

Cuprite forms impressive dark-red crystals in the cubic habit, but the locales in the Southwest rarely yield crystals large enough to facet. More common are cuprite hairs, which are fragile and unsuitable for lapidary work; even large crystals would be difficult to facet, being soft. Luster varies, from submetallic to a sharp, adamantine appearance. Penetration twins are common, and twelve-sided dodecahedrons are rare. The streak is unusual, appearing as a shiny, metallic red or metallic red brown.

There are numerous cuprite locales associated with the various copper districts of the Southwest. Cochise County and Pinal County in Arizona have provided striking specimens in the past, and *Rockhounding Arizona* covers them. *Rockhounding New Mexico* describes the Hanover-Fierro mining district.

# Galena

*Large gray galena cube*
PHOTO COURTESY OF THE RICE NORTHWEST MUSEUM OF ROCKS AND MINERALS

**Lead sulfide, PbS**
**Family:** Metal sulfides
**Mohs:** $2\frac{1}{2}$
**Specific gravity:** 7.4–7.6
**Key test(s):** Gray, soft, stairways
**Likely locale(s):** Known sulfide areas

Galena is the primary ore for lead, and thus specific gravity is one of the key clues for identifying galena in the field. The dull, bluish-gray color, dull luster, and cubic, stair-stepped crystal structure are other easy clues to spot. Finally, galena is relatively soft, and you should be able to scratch it with your fingernail or a piece of calcite. Galena leaves a dark-gray streak. It has a brittle fracture. Cubes are common and unforgettable as well. Galena is usually found in veins and larger disseminated deposits and can mix with argentite, tetrahedrite, or sphalerite to create rich lead, silver, and zinc ores.

Galena occurs in scattered deposits across the Southwest. In Texas Brewster and Culberson Counties contain galena. In Utah Salt Lake County, Utah County, and Emery County host galena to some degree. New Mexico, Nevada, Arizona, and Southern California also contain scattered galena deposits; Arizona's Socorro County has produced fine specimens. *Rockhounding New Mexico* is a good source for collecting locales.

# Goethite

*As goethite continues to oxidize, it becomes a yellow or orange pigment.*
PHOTO COURTESY OF THE RICE NORTHWEST MUSEUM OF ROCKS AND MINERALS

**FeO(OH)**
**Family:** Iron oxides
**Mohs:** 5–5$\frac{1}{2}$
**Specific gravity:** 3.3–4.3
**Key test(s):** Crumbly; iron staining
**Likely locale(s):** Weathered iron mineralization

Goethite is the mineral residue left behind when an iron sulfide such as pyrite loses its sulfur ion. Because it is rich in iron, it serves as a valuable iron ore, and the ancient cave painters used goethite for pigments, calling it ochre. It is often found with limonite, another rusty yellow iron ore. It is typically brown, yellow, or even orange, usually as a mass but sometimes as a coating. The streak shows the same variations. Crystals are rare, but goethite easily forms pseudomorphs as it oxidizes from a sulfur-rich parent.

Goethite is scattered across the Southwest, in every major metal sulfide district, but the guidebooks don't specify any collecting locales.

# Gold

*Gold on quartz from the Sixteen to One Mine, Sierra County, California*

**Au**
**Family:** Elemental metal
**Mohs:** 2½–3
**Specific gravity:** 19.3
**Key test(s):** Weight, color
**Likely locale(s):** Quartz veins

Gold is very dense, at around nineteen grams per cubic centimeter, but it is also soft enough to scratch with calcite. Two minerals have earned the name "fool's gold" because they mimic gold's appearance—pyrite, which is brassy, harder, and smells like sulfur when crushed; and mica, which breaks easily with a knifepoint and tends to float. Gold has a characteristic metallic luster and produces a golden streak if you dare rub it on a streak plate. Raw gold is rarely, if ever, pure. It is usually alloyed with silver, which affects the color significantly. Crystals are exceedingly rare.

California's Mother Lode District is famed for its gold production; elsewhere, many desert locales dot the landscape across Southern California. Most gold panning in Texas is restricted to the Llano Uplift in central Texas. New Mexico has seen placer gold activity in the Ortiz Mountains and lode production in Socorro County. Arizona also has most of its gold produced as a by-product of copper mining, but there has been historic activity in the Oatman District near Wickenburg, Santa Cruz County, and elsewhere. Nevada is a bit dry for a serious placer mining expedition, but the creeks draining the famed Virginia City district are a good option.

# Hematite

*Hematite was once known as "kidney stone" due to its red color and botryoidal habit.*
PHOTO COURTESY OF THE RICE NORTHWEST MUSEUM OF ROCKS AND MINERALS

**Iron oxide, $Fe_2O_3$**
**Family:** Metal oxides
**Mohs:** 5–6
**Specific gravity:** 4.9–5.3
**Key test(s):** Bright red on metal
**Likely locale(s):** Iron-rich mineralized areas

Hematite is rich in iron and is the most common iron ore thanks to dominating banded iron formations. In the field hematite can appear gray or black and has a distinct metallic luster, but after prolonged exposure to air, hematite will eventually start to show its characteristic red signature. Hematite has a striking red streak, which is one telltale sign. Crystals are varied; they can be hexagonal, tabular, or columnar. Hematite can also display a rounded, bubbly botryoidal habit. Limonite, ilmenite, and magnetite are all similar in appearance, but the deep-red streak sets hematite apart from imposters. Hematite can dominate tropical soils and certain clays, concentrating to a point where it is designated as red ochre, one of the oldest color tints known to man. Similarly, yellow ochre is also hematite, but it contains extra water that results in a yellow, not red, color.

Hematite is scattered across the mining districts of the Southwest; *Rockhounding New Mexico* lists multiple opportunities. Excellent specimens have come from La Paz County, Arizona, and the Bouse area is listed in *Rockhounding Arizona*. The Cedar City Iron Mines in Utah still supply hematite samples in the tailings and are mentioned in *Rockhounding Utah*.

# Magnetite

*Magnetite is also called "lodestone" but is uncommon in large specimens.*

**Iron oxide, Fe$_3$O$_4$**
**Family:** Metal oxides
**Mohs:** $5\frac{1}{2}$–$6\frac{1}{2}$
**Specific gravity:** 4.9–5.2
**Key test(s):** Magnetic; heavy, hard, and dark; attracts magnetic dust
**Likely locale(s):** Iron-rich mineralized areas

MINERALS

Magnetite, also known as lodestone, is easy to identify in the field thanks to its magnetic properties. It is usually dark or gray black, with a metallic luster, and leaves a black streak. Crystals are rare and usually small in the isometric habit and typically form stubby, doubly terminated octahedrons. Magnetite is quite common, as it tends to remain behind when many igneous rocks erode. Almost all rivers and streams carry some quantity of magnetite-rich "black sands" in their cracks and underneath bigger rocks. Many prospectors save all of their black sands because, in addition to magnetite, they can usually find palladium, platinum, and other rare-earth metals in the mix. Ocean beaches typically concentrate magnetite thanks to wave action.

Magnetite deposits are scattered throughout the Southwest, in most of the usual metal-mining districts. Arizona's La Paz County is a noted supplier of excellent crystalline specimens. The Magnetite Mine in the Jicarilla District of New Mexico has produced magnetite and hematite, and fine specimens have come from the Hanover-Fierro District of Grant County, New Mexico. Consult *Rockhounding New Mexico* for more information.

# Malachite

*Polished and unpolished malachite from Bisbee, Arizona*
PHOTO COURTESY OF THE RICE NORTHWEST MUSEUM OF ROCKS AND MINERALS

**Copper Carbonate, $Cu_2CO_3(OH)_2$**
**Family:** Carbonates
**Mohs:** $3\frac{1}{2}$–4
**Specific gravity:** 4.1–4.3
**Key test(s):** Green; presence of azurite
**Likely locale(s):** Low-grade copper deposits

This mineral can be a welcome sign in otherwise perplexing or barren outcrops. Even in low concentrations, malachite leaves a telltale green stain, indicating that there is some form of mineralization present. In a pure form malachite makes for a nice specimen and will take a high polish. The key characteristic for identifying malachite is its bright-green appearance. Malachite also has a light-green streak. Malachite often displays banding, especially when prized malachite stalactites are sliced horizontally.

Arizona's copper-mining region is probably the most notable North American locale for malachite, but every mining region known for copper will display at least superficial malachite staining. *Rockhounding New Mexico*, *Rockhounding Arizona*, and *Rockhounding Nevada* are the best places to start to plan a collecting trip.

# Meteorite

*Meteorite fragment from Meteor Crater, Arizona*
PHOTO COURTESY OF THE RICE NORTHWEST MUSEUM OF ROCKS AND MINERALS

**Fe, Ni, and Co (iron); $Fe_2O_3$, MgO, and $SiO_2$ (stony)**
**Family:** Extraterrestrial
**Mohs:** Varies by nickel-iron content
**Specific gravity:** Varies by nickel-iron content
**Key test(s):** Widmanstätten pattern
**Likely locale(s):** Anywhere, but deserts have slower oxidation

There are iron, stony-iron, and stony meteorites. Most meteorites are actually stony, or primarily stone, but they have enough iron to at least attract a magnet on a string. Nickel is also a key component, measuring up to 7 percent. Most clues between meteorites and meteor-wrongs rely on sight. For example, look for the presence of a fusion crust, which is simply evidence that the rock burned its way through the Earth's atmosphere. This will typically appear as a black skin, and if the rock has been broken or fractured, you'll see a bright interior beneath the black crust. Also, look for thumbprint-shaped divots called regmaglypts, which are evidence of the heat generated while passing through the atmosphere.

The best place to scout for meteorites is in a known "strewn field" surrounding an observed fall. Barring that, old lake beds in desert environments are productive because the limited moisture ensures that the iron-rich rocks don't rust and fall apart. Check NASA's database at https://ares.jsc.nasa.gov/meteorite-falls for more information about an area you want to search.

# Molybdenite

*Rare molybdenite crystal*
PHOTO COURTESY OF THE RICE NORTHWEST MUSEUM OF ROCKS AND MINERALS

**Molybdenum sulfide, MoS$_2$**
**Family:** Metal sulfides
**Mohs:** 1–1$\frac{1}{2}$
**Specific gravity:** 4.6–5.1
**Key test(s):** Green streak; gray, soft, greasy
**Likely locale(s):** Interesting mineralized zones

Molybdenite is the chief ore of molybdenum. It is somewhat rare and usually occurs only as small, dull gray lumps. One interesting field test is that molybdenite creates a green streak. When big enough to conduct a scratch test, molybdenite is extremely soft, and a fingernail should scratch it easily. Crystals are rare, hexagonal, and tabular, but easily deformed. Fracture is unlikely, as it is so flexible and soft. Graphite is also lead gray but is lighter than molybdenite. Galena is also gray and soft, but it has a blue-gray streak and cubic crystal structure.

Molybdenite is not a highly sought mineral and doesn't show up much in the guidebooks. San Diego County in Southern California contains several significant molybdenite mines and prospects. Brewster and Presidio Counties in Texas also list multiple mines. Utah contains over a dozen sites, as does Nevada, while Arizona has around 200 locales in its inventory. Yavapai County in Arizona has offered up excellent crystal specimens of molybdenite.

# Platinum

*Platinum nugget in the form of a cube*
PHOTO COURTESY OF THE RICE NORTHWEST MUSEUM OF ROCKS AND MINERALS

**Pt**
**Family:** Elemental metal
**Mohs:** 4–4$\frac{1}{2}$
**Specific gravity:** 21.5
**Key test(s):** Silvery gray and heavy
**Likely locale(s):** Black sands; accessory in metals mining

Platinum is a dull white-gray metal that looks somewhat like silver but is at least twice as dense. Platinum is also far more rare than silver, occurring mostly in the Southwest as an accessory in black sands with other Platinum Group Metals (PGMs), which include ruthenium, rhodium, palladium, osmium, and iridium. Because platinum does not oxidize, it is useful in jewelry, and it has industrial uses as well. Most of the world's platinum comes from South Africa.

Minor platinum showings are common in heavy black sands along the Pacific Coast. Surf action near Lompoc, California, created a minor platinum-rich gold placer there. The Goodsprings District in Nevada reported minor platinum, as did the San Domingo placers in Maricopa County, Arizona. The Horseshoe Bend placers on the Green River in Utah and a string of placers along the upper Colorado River in Utah also reported platinum showings. The Trinity River system in Northern California is a known platinum producer. The Bingham Canyon copper mine in Salt Lake County also reported platinum.

# Pyrite

*Brassy pyrite crystals are distinctive.*
PHOTO COURTESY OF THE RICE NORTHWEST MUSEUM OF ROCKS AND MINERALS

**Iron sulfide, FeS$_2$**
**Family:** Metal sulfides
**Mohs:** 6–6$\frac{1}{2}$
**Specific gravity:** 4.9–5.1
**Key test(s):** Hardness
**Likely locale(s):** Known sulfide areas

Pyrite, or iron pyrite, is also known as fool's gold, thanks to its brassy, yellowish color. Pyrite generally forms in the cubic crystal habit and often has striations, or lines, on its crystal faces. Other times, pyrite crystals tend to twin, interlock, and form interesting masses. There are dozens of varieties of pyrite, with various replacements for the iron ion. Pyrite tarnishes rapidly, becoming darker and somewhat iridescent as oxygen attacks the iron-sulfur bond. Pyrite leaves a brownish-black or even greenish-black streak that smells faintly of sulfur. Pyrite is hard, registering as high as 6$\frac{1}{2}$ on the Mohs scale, just below quartz. Many forms of pyrite are highly collectible.

Pyrite is common in metal-mining districts where sulfides are common. Arizona has yielded "pyrite dollars" south of Tucson, which are sedimentary forms. Elsewhere in Arizona, locales include the Harquahala District in La Paz County; the Kingman District, Orphan District, Cameron District, and Vermillion Hills District, all in Coconino County; and dozens of locales in the Prescott and Tucson areas. Utah, New Mexico, Nevada, and Southern California all contain pyrite-collecting locales associated with precious metals.

# Scheelite

*Orange octahedron of scheelite, from the Cohen Mine, Dos Cabezas Mountains, Cochise County, Arizona*
PHOTO COURTESY OF THE RICE NORTHWEST MUSEUM OF ROCKS AND MINERALS

$CaWO_4$
**Family:** Tungstates
**Mohs:** $4\frac{1}{2}$–5
**Specific gravity:** 5.9–6.1
**Key test(s):** Silvery (when fresh); tarnishes quickly
**Likely locale(s):** Economic ore deposits

Scheelite is the most common ore for tungsten, a metal important in the steel industry. Scheelite is typically yellow to yellow orange but can be white, colorless, light gray, or other pale shades. Pure scheelite fluoresces a nice sky blue under shortwave ultraviolet light. It is marked by a vitreous luster, has good cleavage, and forms excellent crystals that were once used to imitate diamonds. Scheelite can occur in granite pegmatites but is most common in hydrothermal veins where tin or gold is also present.

Scheelite is another metal sulfide common to the mining districts of the Southwest. It isn't a popular collectible, and the guidebooks barely mention it. Southern California contains more than 200 mines and prospects reporting scheelite. Arizona's Tucson area contains at least 100 similar reports.

# Siderite

*Siderite from the Campbell Shaft, near Bisbee in Cochise County, Arizona*
PHOTO COURTESY OF THE RICE NORTHWEST MUSEUM OF ROCKS AND MINERALS

**Iron carbonate, FeCO$_3$**
**Family:** Carbonates
**Mohs:** $3\frac{1}{2}$–$4\frac{1}{2}$
**Specific gravity:** 4
**Key test(s):** Yellow-brown coatings
**Likely locale(s):** Bedded sedimentary deposits

Siderite is typically yellow, brown, or tan. Siderite crystals are rare; instead, siderite usually occurs as yellow-brown masses. It is most commonly found in hydrothermal veins and in bedded sedimentary deposits rich in calcium carbonate, but it also occurs as concretions in shales and sandstones. It is rich in iron by weight and is an important iron ore because it contains little sulfur. Siderite crystals are most common in pegmatites, found as trigonal or hexagonal tabs with vitreous, silky, or pearly luster. It has a white streak.

There are numerous siderite locales across the Southwest, but most collectible specimens come from Colorado.

# Silver

*Native silver from Arizona. Note the pinkish tint of "horn silver" in the sliced sample, lower left.*

**Ag**
**Family:** Elemental metal
**Mohs:** $2\frac{1}{2}$–3
**Specific gravity:** 10.1–11.1
**Key test(s):** Silvery (when fresh); tarnishes quickly
**Likely locale(s):** Economic ore deposits

Silver is actually rare in its native state because it forms oxides such as argentite so readily. Isometric crystals are especially rare and very collectible. Silver has a characteristic shiny, metallic luster when fresh, appearing light gray or whitish gray at times. However, silver tarnishes quickly to black, brown, or yellow. Two key tests are for hardness and specific gravity. Lead forms crystals easier and has the characteristic stair-step pattern. Platinum is heavier than silver and even more rare, but some regions are noted for platinum nuggets, so do your research. Silver specimens almost always have some gold present, forming what the miners called "electrum."

Arizona has more than 100 silver mines and prospects. The classic Arizona collecting locales for good silver specimens are around Bisbee, Arizona. There are numerous locales across Nevada, with the Comstock Lode near Reno being the most famous. Utah lists dozens of silver mines and prospects, including the Horn Silver Mine in Beaver County. Hudspeth County in west Texas has contained some minor silver prospects. The Santa Ana Mountains of Southern California also receive mention in the literature.

# Smithsonite

*Smithsonite from the Kelly Mine, near Magdalena in Socorro County, New Mexico*
PHOTO COURTESY OF THE RICE NORTHWEST MUSEUM OF ROCKS AND MINERALS

**Zinc carbonate, $ZnCO_3$**
**Family:** Carbonates
**Mohs:** $4\frac{1}{2}$
**Specific gravity:** 4.4–4.5
**Key test(s):** Hardness
**Likely locale(s):** Zinc deposits

Smithsonite is prized when found as a translucent green mass but can occur in various colors, including white, gray, pink, and even purple. It rarely occurs in crystalline form, instead presenting in botryoidal masses. It has a white streak, with a vitreous or pearly luster. Smithsonite usually forms in oxidized zones containing other zinc minerals such as sphalerite.

Smithsonite is found throughout the Southwest, with New Mexico hosting some of the most collectible deposits. Check *Rockhounding New Mexico* for likely locales near Magdalena.

# Sphalerite

*Sphalerite crystal on a bed of chalcopyrite and dolomite*
PHOTO COURTESY OF THE RICE NORTHWEST MUSEUM OF ROCKS AND MINERALS

**(Zn,Fe)S**
**Family:** Metal sulfides
**Mohs:** $3\frac{1}{2}$–4
**Specific gravity:** 3.9–4.2
**Key test(s):** Pale-yellow or light-brown streak
**Likely locale(s):** Economic ore deposits

Sphalerite is the primary ore for zinc, and it is often very dark, depending on how much iron is present. It can also be green, yellowish, or red if the iron content is low. Well-formed crystals tend to be resinous or greasy black. Some varieties fluoresce.

Sphalerite is found throughout the Southwest metal-mining districts. Presidio County in Texas hosts several locales, for example. The Empire Zinc Mine in Grant County, New Mexico, has produced fine specimens, as has Graham County in Arizona. There are hundreds of locales reporting sphalerite in Arizona. Consult the FalconGuides for Arizona, New Mexico, and Texas for more information.

# Stibnite

*Fibrous stibnite*
PHOTO COURTESY OF THE RICE NORTHWEST MUSEUM OF ROCKS AND MINERALS

**Antimony sulfide, $Sb_2S_3$**
**Family:** Metal sulfides
**Mohs:** 2
**Specific gravity:** 4.6
**Key test(s):** Soft, slender gray crystals
**Likely locale(s):** Hydrothermal deposits

Stibnite is the principal ore of antimony, and it is highly toxic. It has a metallic-gray luster and streak and is exceedingly soft, so it is easy to crush into a powder. It was known as "kohl" in the ancient Middle East and was used as a cosmetic eye liner, which would not have been healthy. Stibnite forms long, slender crystals, some quite stunning, and it makes an attractive addition to your collection. It is often associated with arsenic minerals such as orpiment, realgar, and arsenopyrite in metal-rich hydrothermal vein systems.

Large deposits of stibnite are rare, but small deposits are scattered across the southwestern states. The Mountain Antimony Mine in Mabey Canyon and the Antimony Peak deposit in Kern County, California, contain deposits. The Getchell Mine in Humboldt County, Nevada, lists associated stibnite. Utah has more than thirty stibnite occurrences.

# Vanadinite

*Vanadinite from Globe, in Gila County, Arizona*
PHOTO COURTESY OF THE RICE NORTHWEST MUSEUM OF ROCKS AND MINERALS

**$Pb_5(VO_4)_3Cl$**
**Family:** Vanadates
**Mohs:** 3
**Specific gravity:** 6.7–7.1
**Key test(s):** Color, heft
**Likely locale(s):** Oxidized zones with galena

Vanadinite is usually a reddish to orange mineral, usually found as small, crystalline masses or coatings. Crystals are hexagonal, but large crystals are rare. Its streak is white to yellow, and its luster is resinous. The presence of lead gives vanadinite a heavy nature, helping distinguish it from mimetite and pyromorphite, but the key distinguishing characteristic is the color. Vanadinite primarily forms in the oxidized zones of lead deposits.

There are numerous vanadinite locales across the Southwest, including Hudspeth County, Texas, and the Groundhog Mine in Grant County, New Mexico. California, Utah, and Nevada all host vanadinite deposits. Excellent specimens come from Arizona, including Pinal County, Pima County, Gila County, and La Paz County.

# Wulfenite

*Red wulfenite from the Red Cloud Mine in the Trigo Mountains of La Paz County, Arizona*
PHOTO COURTESY OF THE RICE NORTHWEST MUSEUM OF ROCKS AND MINERALS

**PbMoO$_4$**
**Family:** Molybdates
**Mohs:** $2\frac{3}{4}$–3
**Specific gravity:** 6.5–7
**Key test(s):** Silvery (when fresh); tarnishes quickly
**Likely locale(s):** Economic ore deposits

Wulfenite contains both lead and molybdenum and thus occurs where galena and molybdenite are also found. It is typically orange, but can occur as red, honey yellow, olive green, or even colorless crystals. It has a resinous luster and white streak and usually occurs as thin, tabular crystals. It can also occur as granular masses. Mineralogists believe wulfenite develops as a secondary mineral, meaning it started out as something else. For example, galena, or lead sulfide, is often found in a hydrothermal vein system, and when oxidized in the presence of molybdenite, wulfenite would crystallize.

There are hundreds of wulfenite locales in Southern California, New Mexico, and Arizona. The Red Cloud Mine in the Trigo Mountains of Arizona is perhaps the most noted producer of fine red wulfenite specimens. The Rowley Mine is listed in *Rockhounding Arizona*. *Rockhounding New Mexico* lists multiple locales.

# PART 3
# GEMS

# Amber (Succinite)

*Amber pendant showing inclusions*

**$C_{12}$, $H_{20}O$ (Tree resin)**
**Family:** Biological
**Mohs:** $2-2\frac{1}{2}$
**Specific gravity:** Variable
**Key test(s):** Cannot dissolve in alcohol
**Likely locale(s):** Coal seams

Amber is not just hardened tree resin. Some of the amber deposits in Central America are not actually old enough to be truly termed an amber—they will still partially dissolve or soften in solvents such as alcohol, acetone, or even gasoline. The name for such specimens is copal. To truly grade as amber, material must undergo more of a change than just losing moisture and volatile chemicals. Instead, it must polymerize—the process of converting smaller molecules into larger ones. This typically takes millions of years to accomplish.

There are reports of amber found in Texas at Woodward Ranch, noted for fluorescents, agates, and other material. Two other descriptions include coal seams along Terlingua Creek and in Cretaceous coal seams at Eagle Pass, all in Brewster County, Texas. The northeast side of Simi Valley in Ventura County, California, also contains one recorded deposit.

# Ammolite

*Raw ammolite (left) and finished jewelry*
PHOTO (LEFT) COURTESY OF THE GEMOLOGICAL INSTITUTE OF AMERICA

$CaCO_3$
**Family:** Carbonates
**Mohs:** $4\frac{1}{2}$–$5\frac{1}{2}$
**Specific gravity:** Variable
**Key test(s):** Iridescence
**Likely locale(s):** Ammonite fossil locales

Ammolite is a vivid, multicolored variety of ammonite fossil that has not only preserved but enhanced the nacre, or colored shell. If you've ever seen an abalone shell, for example, you know the beautiful colors that display. Ammolite is fossilized ammonite shell that is prized by jewelers for the iridescent play of colors possible. Striking orange, green, red, yellow, and blue colors all vie for attention in ammolite. Since ammolite is often very thin and quite fragile, it must usually be layered with harder material for protection, and thus is a symbol of lapidary skill.

Most of the world's ammolite supply originates in Canada. However, equivalent rocks to the Cretaceous Bearpaw Formation of Canada extend into the southwestern United States, and a significant deposit was discovered in central Utah among ammonite fossils there. Consult mindat.org for more information.

# Aquamarine

*Cluster of aquamarine crystals*
PHOTO COURTESY OF THE RICE NORTHWEST MUSEUM OF ROCKS AND MINERALS

**$Be_3Al_2(SiO_3)_6$**
**Family:** Beryl
**Mohs:** $7\frac{1}{2}$–8
**Specific gravity:** 2.8
**Key test(s):** Hardness
**Likely locale(s):** Granite pegmatites

Aquamarine is another gem variety of beryl, like emerald, heliodor, morganite, red beryl, and goshenite. Aquamarine is blue, and when deep, striking blue, it is highly sought as a gemstone. It is quite rare, thus worth prospecting for. The hardness of around 8 makes it hard to take a streak, but it is white. The luster is vitreous to resinous, cleavage is poor, and it does not fluoresce. It is usually associated with granite pegmatites.

The Southwest states contain a few scattered locales. In Arizona there are occurrences at the Thompson Beryl Claims in the Swisshelm Mountains of Cochise County, and at the Monte Cristo pegmatite at Weaver Peak in Yavapai County. In Utah there are two deposits noted in Tooele County—in the Ibapah Mountains and at Elephant Knoll. In Southern California there are occurrences in the Jacumba District, the Ramona District, the Rincon District, the Warner Springs District, the Cahuilla District, and the Pala District, with the Pala mines hosting fee-dig opportunities.

# Diamond

*Small "raw" diamonds, unfaceted*
PHOTO COURTESY OF THE RICE NORTHWEST MUSEUM OF ROCKS AND MINERALS

**Carbon, C**
**Family:** Element
**Mohs:** 10
**Specific gravity:** 3.5
**Key test(s):** Hardness
**Likely locale(s):** Associated strictly with kimberlite pipes

Diamond specimens are clear, yellow, pink, blue, purple, brown, and even black. The hardness test is the best indicator—pure, strongly crystallized diamond, at 10 on the Mohs scale, scratches everything. Another good indicator is that diamond has an adamantine, or outstanding, luster. Its crystal habit is octahedral, and cleavage is perfect in four directions, which is rare. Clear agate, and clear gem-quality quartz such as Herkimer "diamonds," look similar to actual diamonds but are hexagonal and won't cleave.

In California, miners found small diamonds in the Trinity River system at Devil's Gate, on Hatfield Creek in the Ramona District in San Diego County, and on Alpine Creek in Tulare County. The Arizona Geological Survey reports an occurrence of graphite and diamonds in Coconino County, suspiciously close to Meteor Crater and possibly of extraterrestrial origin.

# Elbaite

*"Watermelon" tourmaline (elbaite) from the Pala District of Southern California*
PHOTO COURTESY OF THE RICE NORTHWEST MUSEUM OF ROCKS AND MINERALS

$Na(LiAl)_3Al_6Si_6O_{18}(BO_3)_3(OH)_4$
**Family:** Silicates
**Hard:** $7-7\frac{1}{2}$
**Specific gravity:** 3.0–3.3
**Key test(s):** Harder than apatite
**Likely locale(s):** Pegmatites

Tourmaline comes in a variety of forms, with the most common being black schorl. That's by far the most common tourmaline species, but semiprecious gem varieties include elbaite, first noted on Elba Island but now primarily known from Southern California; indicolite, which is blue; rubellite, which can be pink or red; and dravite, which is brown. Tourmaline has a vitreous luster, leaves a white streak, and forms a hexagonal crystal with no cleavage.

Gem-quality tourmalines are usually found in vugs and cavities within granite pegmatites, such as in Southern California. The Pala District offers interesting fee-dig opportunities.

# Garnet

*Two varieties of garnet: spessartine (left) and hessonite (right)*
PHOTOS COURTESY OF THE RICE NORTHWEST MUSEUM OF ROCKS AND MINERALS

**$X_3Y_2(SiO_4)_3$ Almandine**
**Family:** Metal silicates
**Mohs:** $6\frac{1}{2}$–$7\frac{1}{2}$
**Specific gravity:** 3.6–4.3
**Key test(s):** Hackly fracture, hardness
**Likely locale(s):** Schist; black sands

Increasing metamorphism ⟶

| Chlorite | Biotite | Garnet | Staurolite | Kyanite | Sillimanite |
|----------|---------|--------|------------|---------|-------------|

Taken as a family, garnet crystals show a vitreous luster, have no streak to speak of, and do not cleave. Garnet is harder than apatite, does not fluoresce like zircon, and has higher specific gravity than tourmaline, plus it is usually associated with schist. That rule isn't hard-and-fast, however, as some gem garnets are associated with granitic pegmatites. Gem-quality garnet is somewhat rare, and adding further value is a tendency for impurities to line up and create four-sided or six-sided stars.

There are at least ninety garnet locales in Southern California, thirty in Nevada, sixty in Arizona, and scattered deposits in west Texas, Utah, and New Mexico. The field guides for each of those states contain locales to explore. Garnet Hill in California has been well known for years. *Rockhounding Nevada* lists the free Garnet Hill dig site, near Ely.

# Jade

*Jadeite specimen, showing rind*
PHOTO COURTESY OF THE RICE NORTHWEST MUSEUM OF ROCKS AND MINERALS

**Manganese silicate, MnSiO₃**
**Family:** Silicates
**Mohs:** $6\frac{1}{2}$–7 for jadeite; $5\frac{1}{2}$–6 for nephrite
**Specific gravity:** 2.9–3.1
**Key test(s):** Can't be scratched by a knife; botryoidal
**Likely locale(s):** Mafic rocks

There are two main varieties of jade. Both are amazingly strong due to interlocking crystals that form nearly unbreakable bonds. First, there is classic jadeite, a sodium-rich, aluminum-rich pyroxene, found mostly in Burma and favored by Chinese rulers for centuries. Jadeite is not quite as hard as quartz. Second, there is nephrite jade, a type of amphibole, also famous in China, but more commonly associated with North American deposits in Wyoming, British Columbia, Washington State, and California. Nephrite is slightly softer than jadeite and usually occurs only as white or shades of green. A high-quality steel knife blade cannot scratch either jadeite or nephrite. Jadeite, which can be purple, blue, lavender, pink, or vivid green, is the more highly prized, but that isn't a hard-and-fast rule, as rare, "mutton-fat" white nephrite jade commands fabulous prices.

The Southwest does not have many jade locales. California is the exception, with jadeite reported near Paso Robles, near Cayacos, and at Jade Cove south of Big Sur, which is described in *Rockhounding California*. The book also describes additional jadeite locales near Covelo on the Eel River.

# Peridot

*Beautiful faceted peridot*
PHOTO COURTESY OF THE RICE NORTHWEST MUSEUM OF ROCKS AND MINERALS

**$Mg_2SiO_4$**
**Family:** Silicates
**Mohs:** $6\frac{1}{2}$–7
**Specific gravity:** 3.2–4.3
**Key test(s):** Olive-green color
**Likely locale(s):** Olivine-rich basalt flows

Peridot is a gem-quality olivine crystal, and while olivine is quite common, large, facetable peridot crystals are rare. Olivine is common in mafic and ultramafic rocks that are rich in iron and magnesium, so expect to find peridot in similar locales. Peridot has a white streak, a glassy luster, and poor cleavage. It is easy to confuse with emeralds. Some rare pallasite meteorites include olivine and peridot and command a premium price.

Peridot is rare in the Southwest, but the most productive locale in North America has been a band of black basalt at Peridot Mesa on Arizona's San Carlos Reservation. This area is closed to collectors. The volcanic maars at Kilbourne Hole, New Mexico, are known for olivine-rich nodules carried from the upper mantle to the surface via basalt flows and ejected as bombs. Some of these bombs contain facetable peridot.

# Precious Opal

*Famed "Freida's Log"—an opalized log from the Royal Peacock Mine, Nevada*
SPECIMEN COURTESY OF JULIE WILSON. PHOTO COURTESY OF GLENN OTTO/OTTO GRAPHICS.

$SiO_2 \cdot nH_2O$
**Family:** Silicates
**Mohs:** $5\frac{1}{2}$–$6\frac{1}{2}$
**Specific gravity:** 2.0–2.2
**Key test(s):** Play of color; luster
**Likely locale(s):** Tuff deposits

Precious opal is a rare form of common opal that owes its stunning color to the way its silica spheres are stacked and packed, diffracting the light and causing color interplay. Precious opal is relatively soft, has no crystal structure and no cleavage, and is usually found in veins and nodules. The streak is white. Precious opal comes in a variety of forms: standard precious opal, displaying all colors of the rainbow, usually after replacing wood; fire opal, usually red or orange, derived from thin seams between lava flows; black opal, very dark in color but with a nice play of colors; and white opal, which is usually white but also exhibits a full play of colors.

The most famous precious opal locale in the Southwest is in the Virgin Valley area of northern Nevada, where the precious opal replaces wood and sometimes bone. There are several fee-dig operations there, such as the Royal Peacock Opal Mine and the Bonanza Opal Mines, described in *Rockhounding Nevada*. There is also a honey opal from Red Rock Canyon in Kern County, California.

# Ruby

*Hexagonal ruby crystals*
PHOTO COURTESY OF THE RICE NORTHWEST MUSEUM OF ROCKS AND MINERALS

**Aluminum oxide, Al$_2$O$_3$**
**Family:** Oxides
**Mohs:** 9
**Specific gravity:** 3.9–4.0
**Key test(s):** Hardness
**Likely locale(s):** Pegmatites, dikes, and metamorphic zones

Rubies are the pink or red variety of corundum, with the presence of chromium spurring the red color. If a ruby isn't red enough, it is sometimes designated as a pink sapphire; the line can be blurry, however. The important point for rubies is that the redder, the better. Treatments and enhancements are very straightforward with rubies, and gemologists look for the natural inclusion of rutile to determine if a ruby has been heated or otherwise treated. Occasionally, rubies have enough rutile needles to develop asterism, or a "star" when polished in a round shape and viewed under light.

There are two known ruby-producing areas in the Southwest. In Utah the Jim Fisk Mine sits on the Sacramento Dike in the Ophir District, located in the Oquirrh Mountains of Tooele County. This was a major lead and silver producer in its day. The second locale is at Cascade Canyon in the San Gabriel Mountains of San Bernardino County, California, where collectors have washed pink rubies from the gravels of the nearby creek.

# Sapphire

*Crystals and water-worn fragments of sapphire*
PHOTO COURTESY OF THE RICE NORTHWEST MUSEUM OF ROCKS AND MINERALS

**Aluminum oxide, $Al_2O_3$**

**Family:** Oxides

**Mohs:** 9

**Specific gravity:** 3.9–4.0

**Key test(s):** Hardness

**Likely locale(s):** Pegmatites, dikes, and metamorphic zones

Sapphire is another well-known variety of gem-quality corundum. Unlike rubies, which are typically red, a sapphire is usually blue, although some sapphires are clear, light gray, or even dark gray. The blue color comes from iron and titanium impurities that affect ion charges, color absorption . . . it all gets very technical. The inclusion of rutile needles results in a "star" effect for properly polished sapphires and is highly desirable. Note that sapphires are one of the easiest gems to treat with heat and enhance or change the color. Sapphires are typically associated with pegmatites and can be difficult to separate from host rock.

There are three sapphire-producing areas in the Southwest, all in California: the Andalusite pegmatite group at Bautista Canyon in Riverside County; the Hazen Corundum prospect in the San Jacinto Mountains of Riverside County; and Cascade Canyon, in the San Gabriel Mountains of San Bernardino County.

# Sunstones

*Sunstones are a variety of feldspar.*

$(Ca,Na)((Al, Sl)_2Si_2O_8)$
**Family:** Feldspars
**Mohs:** $6-6\frac{1}{2}$
**Specific gravity:** 2.6
**Key test(s):** Shape, color, hardness
**Likely locale(s):** Basalt flows

Sunstones are a form of plagioclase feldspar, making them yet another incarnation of a sodium calcium aluminum silicate. They are also known as labradorite and are usually yellow, but the inclusion of tiny copper crystals provides a unique "Schiller" effect that can be deep red and makes for an interesting twinkle as the specimen rotates. Sunstones form in basalt vesicles and are usually very fractured and irregular. The euhedral crystals have a vitreous luster and a white streak and cleave in two directions. They don't look anything like the orbs and blebs of opal- or agate-filled basalt vesicles and should be scratched by quartz. Sunstones take a nice polish.

Small sunstones litter the ground at Sunstone Knoll near Delta, Utah. The famed Woodward Ranch in Brewster County, Texas, contains labradorite. New Mexico's Sierra County hosts a pair of labradorite deposits. The Crestmore quarries in Riverside County, California, contain labradorite. Arizona hosts several labradorite deposits, mostly in the Tucson area.

# Topaz

*Large topaz cluster chunk (left), with small, twinned crystal from Utah's Topaz Mountain (right)*
PHOTO (LEFT) COURTESY OF THE RICE NORTHWEST MUSEUM OF ROCKS AND MINERALS

**Aluminum silicate, $Al_2SiO_4(F,OH)_2$**
**Family:** Silicates
**Mohs:** 8
**Specific gravity:** 3.4–3.6
**Key test(s):** Pink (when fresh), hard
**Likely locale(s):** Rhyolite flows

Topaz is an interesting mineral that doesn't seem to get much respect for its natural state. It is commonly irradiated or coated to produce more pleasing results. Typically, topaz is clear when pure, but impurities can cause it to appear white, light gray, or even pink or yellow. Laboratory treatments result in more stunning blues and orange colors. Topaz has a glassy to vitreous luster and forms stubby, prismatic crystals in the orthorhombic crystal system, with excellent terminations. The best test in the field is its hardness—it will scratch quartz. Cleavage is excellent in one direction, and striations are common lengthwise on crystal faces, but topaz is often found in massive lumps. Topaz forms at high temperatures, among silica-rich igneous rocks such as rhyolite, or in cavities within granite pegmatites.

Topaz is the state gem of Utah, and it is found in Beaver, Juab, and Tooele Counties there. The most prominent topaz-collecting locale in the Southwest is at Topaz Mountain, covered in *Rockhounding Utah*. Located in the Thomas Range, the topaz crystals formed in vugs and cavities within a rhyolite. The only topaz deposit in Texas is in Mason County, in the middle of the state.

# Turquoise

*Polished turquoise nodule and cabochon from Bisbee, Arizona*
PHOTO COURTESY OF THE RICE NORTHWEST MUSEUM OF ROCKS AND MINERALS

$CuAl_6(PO_4)_4(OH)_8 \cdot 4\text{-}5H_2O$
**Family:** Phosphates
**Mohs:** 5–7
**Specific gravity:** 3.5–4.0
**Key test(s):** Blue; waxy luster
**Likely locale(s):** Veins, seam fillings; near copper mines

Turquoise is practically synonymous with Southwest Native American jewelry. Sadly, because demand is so high, some turquoise is dyed or impregnated to enhance its colors and to bring out more of the coveted blue. Natural turquoise rarely forms crystals, occurring instead as nuggets and filled-in fractures in the host rock. Turquoise has a waxy luster and a faint, bluish-white streak, and the powder is soluble in hydrochloric acid. Under long-wave UV light, turquoise may fluoresce. Black limonite veining is common. Some of the minerals confused with turquoise include chrysocolla, which is much softer, and variscite, which is usually greener and softer.

There are numerous turquoise deposits in the literature across the Southwest, with at least forty in Arizona alone. Famous locales include the area around Cerillos, New Mexico, one of the oldest known producers; a pair of high-quality mines in Arizona; and a long belt of turquoise across Nevada. California hosts turquoise at Lane Mountain and in the Turquoise Mountains in San Bernardino County, among other places. The famed fee-dig operation hosted by the Otteson Brothers outside Beatty, Nevada, is a family-friendly opportunity.

GEMS

# Glossary

**alluvium:** Dirt, usually. Stream and river deposits of sand, mud, rock, and other material. Sometimes sorted, if laid down in deep water; otherwise can be unsorted, if deposited during floods, earthquakes, etc. If glaciers were involved, the term "till" is used.

**anthracite:** The hardest and most intensely metamorphosed form of coal.

**arkose:** Sandstone that has a lot of unsorted, broken-up pieces of feldspar and quartz. Usually hard and not easily eroded.

**basement:** The "lowest" and oldest rocks around, usually metamorphic, and frequently dating to the Precambrian or Paleozoic ages. They are usually less prone to erosion and make up mountain ranges and stunning cliffs. "Basement" refers to their placement at the bottom of a stratigraphic table.

**batholith:** General term that refers to extremely large masses of coarse intrusive rock such as granite. Any rock formation over 100 km$^2$ (39 square miles) is considered a batholith.

**bedding:** The tendency of sedimentary rocks such as sandstone to reside in visible zones or marker beds, similar to tree rings.

**bleb:** A round or oval cavity, air bubble, hole, or vesicle, usually in basalt, and sometimes filled with opal, agate, or chalcedony, and then eroded out.

**chemical sediment:** Refers to the way certain limestones and dolomites precipitate material such as calcium carbonate, which falls to the bottom of the sea or bay and accumulates.

**clasts:** Catchall term for the clay, silt, sand, gravel, cobbles, and boulders that make up nonchemical sedimentary rocks. The size of the clasts then determines the name of the rock.

**clay:** Usually refers to the smallest mineral fragments, smaller than 2μm or $\frac{1}{255}$ mm.

**cobble:** Fancy term for rocks between a pebble and a boulder. The exact definition of a cobble is anything from 64 to 256 millimeters in size.

**contact metamorphism:** The result of a hot igneous intrusion on the country rock. The contact zones between the intrusion and the surrounding rock can sometimes house interesting mineralization.

**density:** Defines the weight per an agreed unit of volume. By weighing the sample and then dunking it in water and measuring the volume of water displaced, we get the density measured in grams per cubic centimeter. The general term "heft" refers to a field test for how dense a typical-size hand specimen feels.

**diatomite:** Usually a white, chalky deposit that, upon microscopic inspection, turns out to be composed of tiny diatom fossils. These beds can often host common opal, precious opal, and zeolite deposits.

**drift:** General term for glacial deposits composed of jumbled debris. Outwash plains and terraces are usually sorted, whereas tills and moraines are unsorted.

**dry wash:** The sign of a seasonal stream that dries up during the summer months. These can be interesting for rockhounds, as specimen sizes are usually bigger because they haven't been severely eroded.

**eolian:** General term for wind deposits such as loess, sand sheets, ripples, and dunes, but also can refer to wind processes such as dust storms, sandblasting, and desert varnish.

**eon:** The longest division of geologic time is the super-eon. The model is: super-eon —> eon —> era —> period —> epoch —> age. Thus, we are in the Holocene epoch of the Quaternary period of the Cenozoic age of the Phanerozoic eon of the Cambrian super-eon.

**epoch:** Shorter subdivision of a geologic period, usually corresponding to observed stratigraphy in the field.

**era:** The four main geologic eras, from oldest to youngest, are the Precambrian, Paleozoic (which starts with the Cambrian), Mesozoic (the age of dinosaurs), and Cenozoic (ours).

**erosion:** The forces and processes that continually grind down mountains and move their debris downwind or downhill.

**evaporite:** As bodies of water dry up under desert conditions, they frequently get white or light-brown crystalline rings around the edges. These minerals are usually pure salt (halite, or sodium chloride) or a related halide, plus borax or gypsum also.

**exfoliation:** This term refers to the way rocks, and in particular granite, tend to slough off skins or layers of outer rock, like an onion. The result is usually a rough, rounded shape, rather than angles and edges.

**facies:** Field term used to describe how sedimentary rocks can be identified by the way they were deposited. The term biofacies describes the distinct fossil assemblage, while the term lithofacies could describe the similarities in a rock's clast size. There could be several distinct facies identifiable in the field that make up an overall formation.

**float:** Describes the difference between rock samples hammered from an outcrop, and thus with a known origin, and samples that exist as cobbles or boulders and not attached to bedrock. Prospectors are able to trace float to its source outcrop.

**flood basalt:** Refers to the way basalt tends to pour out of cracks and vents and form rivers of liquid rock, and thus create plateaus of flat, layered deposits. By comparison, andesites pile up.

**flow cleavage:** Describes the tendency of metamorphic rocks to arrange flat, elongated crystals into a parallel structure.

**formation:** This is a key term to understand in field geology. Geologists assign formation names to mappable, recognizable rock assemblages and also note a "type" locale that defines the rest of the unit. Formations can be lumped together into groups or even supergroups. To become a formation, a group of similar rocks must be big enough to be worth the bother, must be the same age, must share some key similarity, and must be traceable across the surface. Some formations are divided into members.

**geologic cycle:** The continuous cycle of destruction, recycling, and rebirth that defines the way the Earth's crust works. There are countless variations on the scenario, but, in general, rocks are created, such as by a volcano, eroded, built back into sedimentary deposits, subducted, cooked, melted, then turned back into lava, and erupted again.

**graded bedding:** Sedimentary rock term to describe the way a creek or river deposits alluviam, pebbles, and boulders in a typical sequence, from coarse to fine. Look for coarse conglomerates at the bottom and fine siltstones at the top.

**gravel:** The best place for pebble pups and rockhounds to search for interesting material. Gravels usually consist of pebbles, cobbles, and boulders, in various ratios, and also contain varying amounts of sand and silt. Gravel bars refresh with each season and provide clues to the surrounding geology.

**hydrothermal vein:** Best spot to investigate for interesting minerals. These hot, chemical-rich solutions, usually quartz, can either find an existing crack in country rock or create their own. If they cool slowly enough, hydrothermal veins can create ore deposits with large crystals prized by collectors.

**ignimbrite:** The igneous rock created when hot volcanic ash and breccias pour out of a volcano or vent and are too heavy to drift away as an ash cloud.

**intrusion:** Catchall term for the various granites, diorites, and related rocks that bulldoze their way through the Earth's crust but never reach the surface. Cooling in place quickly results in fine-grained material; cooling slowly gives the individual elements more time to build up into larger crystals.

**laccolith:** This is a small intrusion that squeezes in horizontally between beds and builds itself out laterally. These deposits usually have a neck, where material fed in, and a dome, depending on how forceful the intrusion was.

**lahar:** When volcanic eruptions mix with melted glaciers, lakes, and snow, the result is a dangerous mudflow called a lahar. The material can flow quickly, but it will soon solidify into a concrete-hard mass of jumbled-up ash, fragments, and pumice.

**lava:** Catchall term for extruded molten rock that includes basalt, andesite, rhyolite, dacite, and others.

**lode:** The prized zone of rich, extended mineralization that usually ensures a successful mining operation. The term is reserved for larger vein networks that cover significant ground.

**luster:** Term for the visual appearance of a mineral's lighted surface. The way minerals reflect light can be helpful for identification, but terms such as metallic and waxy are not completely standardized.

**mafic mineral:** These are the dark, heavy minerals that are rich in iron and magnesium, such as pyroxenes, amphiboles, and olivines.

**magma:** The molten lava that eventually forms igneous rock when it cools. Magma that cools without eruption is called an intrusion; otherwise it is extrusive.

**magma chamber:** The source cavity or reservoir for magma traveling up through the Earth's crust.

**mass spectrometer:** The one instrument you wish you had, because it can count ions and provide their exact distribution. A cheap, handheld mass spectrometer would revolutionize field geology.

**Mohs scale of hardness:** The observation-based method of ranking a mineral by what it can scratch and what, in turn, can scratch it. Thus, diamond is alone at the top of the list at 10, and it can scratch corundum, which can scratch topaz, which can scratch quartz, at 7. The remaining minerals are: orthoclase feldspar, 6; apatite, 5; fluorite, 4; calcite, 3; gypsum, 2, and talc, the softest, at 1.

**native metal:** Metals in their purest form are not significantly combined with oxides, sulfides, carbonates, or silicates, and are thus native. Gold, silver, copper, platinum, and mercury are examples.

**oil shale:** A dark, organic-rich shale that sometimes contains enough petroleum-based ingredients to burn.

**oolite:** Refers to the small, round form taken when calcium carbonates start to coat sand grains and roll around in a lime-rich sea bottom.

**ore:** The term used to describe a viable mineral deposit that is worth mining. Usually refers to a metal-based mineral that must be milled.

**Original Horizontality:** This term refers to the idea proposed by Nicholas Steno (1638–86) that sedimentary rocks are laid down flat. Since many sandstone beds are currently tilted, this simple concept meant other forces were at work.

**outcrop:** A cliff, ledge, or other visible clue to the rock formations below.

**pegmatite:** A key igneous rock, usually found as a vein or dike, with very large grains of mica, feldspar, and tourmaline, among others. Pegmatites tend to form cavities and vugs where large crystals can accumulate without crowding into each other and damaging their crystal structure. Pegmatites sometimes host smoky quartz, beryl, topaz, aquamarine, and other exotic gems.

**pelagic sediment:** Catchall term for the fine sediments that slowly accumulate in deep marine environments. Rather than relying on silt or clastic debris, this material is predominantly derived from shells of microscopic organizations such as foraminifera. Accumulation rates are as slow as 0.1 cm per 1,000 years.

**porphyry:** a hard igneous rock containing crystals, usually of feldspar, in a fine-grained, dark red or purple matrix.

**reaction series:** Refers to the behavior of a cooling magma where some minerals form at high temperature, and others as temperatures cool. These conditions are observable in a laboratory setting.

**regolith:** Another catchall term, this time describing the various rock fragments and erosional debris that lie on bedrock, including alluvium, clastics, and others.

**replacement deposit:** Describes a particular type of ore deposit where hot, circulating solutions first dissolve a mineral to form a cavity, then fill the void with a new material.

**sedimentary structure:** Describes the various relicts of a sedimentary rock's deposition, such as ripples, cracks, bedding, zones, layers, etc.

**stratification:** The tendency of sedimentary rocks to form in flat, parallel sequences that are mappable at the surface for considerable distances. Geologists identify patterns and unravel the sequence of events represented in the strata.

**stratigraphic column:** Stratigraphy is the science and study of sedimentary rock outcrops. Stratigraphers draw stratigraphic columns that pictorially represent the measured or inferred relations of rock outcrops, especially age. Metamorphic and igneous rocks occasionally show up at the bottom in stratigraphic columns, as it is primarily a tool to understand sedimentary rocks.

**streak:** Refers to the color of the powdered mineral dust left behind when a mineral is scraped across a streak plate. The color of this fine rock powder is a more true reflection of the mineral than visual appearance. Since a streak plate is about 7 on the Mohs hardness scale, only minerals less than 7 can easily be tested by streak.

**Superposition, Principle of:** Another theory traced to Nicolas Steno (1638–86), who pointed out that a rock formation or strata that sits on top of another distinct layer must usually be younger than the rock below. The older rock will always be at the bottom, except in rare conditions.

**talus:** The impressive accumulation of debris below a cliff or prominent outcrop. Because the rocks are shifting constantly and continuing to accumulate, few plants can get a foothold. Also called scree.

**tectonics:** Geologic theories of how the Earth's crust continually moves and shapes new rocks. To compare planets, on Mars there is little evidence of tectonics, and there is basically one volcano, Olympus Mons, rising 13 miles (21 km) above the planet surface. On Earth the plates continually shift, resulting in arcs, troughs, and collision zones.

**texture:** Describes a rock's grain size, crystal size, if the grains are uniform or variable, if the grains are rounded or angular, and if there is any evidence of orientation to the grains.

**till:** Jumbled mess of glacial debris, with little to no recognizable bedding present and sediment sizes ranging from rock flour and clay all the way to massive boulders.

**tuff:** The term used to describe ash, pumice, volcanic breccias, and other debris. Some ash beds yield petrified wood and fossil leaves; others are welded solid.

**ultramafic rock:** Igneous rocks such as dunite, peridotite, amphibolite, and pyroxenite that consist of primarily mafic minerals and have less than 10 percent feldspar.

**volcanic ash:** The fine rock fragments and glassy, angular material ejected from a volcano into the air.

**volcanic breccia:** A pyroclastic rock made up of angular fragments that show little or no sign of waterborne movement. Particle sizes are greater than 2 mm in diameter.

**xenolith:** Literally, "foreign rock" where a piece of country rock is picked off the walls or otherwise incorporated into a rising or spreading dike or intrusion.

**zeolite:** Common aluminosilicates formed in volcanic rocks, such as basalt, where alkaline groundwaters circulate at low temperatures and create a ringed "molecular sieve" structure.

# Index

# About the Author

Garret Romaine, an avid rockhound, prospector, fossil collector, and gem hunter, is the author of numerous rockhounding field guides and handbooks. Romaine is a retired associate professor in the English Department at Portland State University and was a columnist for *Gold Prospectors* magazine for fifteen years. He is a former executive director of the Rice Northwest Museum of Rocks and Minerals.